THE EVOLUTION OF LOCOMOTIVE VALVE GEARS

by

T. H. Shields

ISBN 1 85761 109 8
First published 1943
© TEE Publishing 1999

The Evolution of Locomotive Valve Gears

Paper read before the Institution by T. H. SHIELDS, Member, on 23rd September, 1943, in London.

The almost universal application of the century old Stephenson's Link Motion and Walschaerts' Valve Gear gives little indication of the amount of ingenuity spent on the numerous mechanisms invented or proposed, for operating the slide valves of the locomotive. Initially plug, tappet or slide valves controlled only the two phases of admission and release of steam to and from the cylinders; however, as the expansive properties of steam became better understood, slide valves were constructed with outside and inside lap, and valve gears were evolved to govern the now common four phase cycle of admission, expansion, exhaustion and compression, and also to reverse the motion of the engine.

Valve gears have a threefold purpose to perform:—
(1) To give the valve its required motion relative to that of the piston.
(2) To vary the travel of the valve.
(3) To reverse its position relative to that of the piston.

The crude engines of the 17th century by the Marquis of Worcester, Savery and Moreland did not employ the use of a piston and the various cocks were manual operated. With the introduction of Newcomen's engine in the first decade of the 18th century earlier attempts to lift water by the condensation of steam were abandoned. Newcomen was the first, at least in England, to use a steam tight piston in a cylinder, and in 1712 he set his engine to work at Wolverhampton, and in course of time pump valves, clacks and buckets were introduced. Humphrey Potter, a boy employed as a " valve turner," added to the engine what he termed " scoggan," this being an arrangement of cords and strings connected to the beam of the engine to automatically open and shut its own valves and cocks. By this arrangement the piston strokes were increased from 10 to 16 per minute, and from this crude device valve gearing originated.

Henry Beighton, of Newcastle, had an engine built in 1718, in which the strings and catches of Potter were taken away and the beam was directly made to actuate the valves. In 1769 Smeaton made a valve gear similar to that of Beighton, this being improved in 1775.

James Watt worked on the idea of the separate condenser for Newcomen's engine in 1765 and obtained his first patent in 1769, his second patent for a double acting steam engine being taken out in 1780. Fig. 1 shows Watt's valve gear for his double acting steam engine built in 1788; on the air pump rod, called the plug tree, were two projecting plugs of wood to operate the valves. As the pump rod moved up and down the plugs came in contact with the handles of the working gear, opening and shutting the valves in turn, the valves being lifted off their seat by a rack and pinion. William Murdock made later improvements on valve gears by the introduction of what is now known as the Cornish Valve Gear.

Up to this period valves of atmospheric and steam engines were of the conical or drop type, but in 1784 Murdock introduced the long D slide valve, in the form of a long double cylindrical valve, the now common short slide valve being due to Matthew Murray in 1802.

Leupold, in Cassel, Germany, published a book in 1712 called " Theatrum Machinarum" in which he described a single acting high pressure steam engine using two pistons, the cylinders being placed above a boiler, and although there is no record of this engine being constructed, James Watt, during 1769-70, studied German with a Swiss-German dyer in Glasgow with the sole purpose of studying Leupold's book.

Taking Leupold's engine for a model, Nicholas Cugnot built a steam carriage in Paris during 1769. The engine worked a single wheel in front of the vehicle, one cylinder on each side, the extension of the piston rods operating ratchets to obtain the rotary motion of the wheel, the steam acting alternately on the pistons being admitted and discharged by a four-way cock. William Murdoch made a small model of a steam carriage in 1784 with a cylinder $\frac{3}{4}$ in. diameter and 3 inch stroke, to which he applied a form of slide valve. In America, Oliver Evans worked on the idea of applying the steam engine as the motive power for wheeled carriages from as early as 1772, and William Symington exhibited a model of a steam carriage in Edinburgh in 1786. Trevithick's

THE EVOLUTION OF LOCOMOTIVE VALVE GEARS

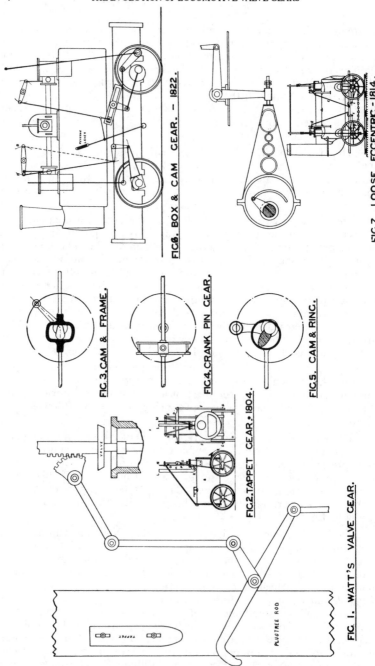

first self-propelled vehicle appeared in 1802, and was likewise operated on the common road.

Fig. 2 illustrates a Trevithick engine of the 1802-04 period.[1] In the cast iron boiler B was fixed a cylinder, at the top of the piston rod C there was a cross beam D, the ends of the cross beam being kept parallel by the vertical rods E, this also preventing the piston from bending. From the ends of D were fixed the connecting rods F which engaged the crank pins G on the front wheels on each side of the locomotive. When the piston was raised or lowered the wheels turned in the direction of the arrow. To operate the ordinary three-way cock controlling the admission and egress of steam to and from the cylinder there was a projection at M on the cross beam D, which acted upon two similar projections on the valve rod N. As the cross beam D rose and fell it struck either at O on its underside or P on its upper side, and this movement of the valve rod N opened or closed the valve R. The steam was exhausted through the waste steam pipe L into the atmosphere. After this date some locomotives had the cocks, controlling the admission and exhaustion of steam from the cylinders, opened and shut by contact with an arm on engine crosshead.

The eccentric is supposed to have been in use on pumping engines previous to Murdock using this contrivance in 1798. However, till about 1820 cam and frame gear, stop valve gear and narrow cam and ring gear were employed to convert the rotary motion of the shaft into the required reciprocating motion of the slide valve. Early Killingworth engines employed the square box and cam gear, these giving fair results in fore gear, but were inferior for back gear working. The cam and ring gear was also called the barrel and cam gear, this being a cam enclosed in a ring and may be considered as the forerunner of the eccentric valve motion, these arrangements being shown in Figs. 3, 4 and 5.

Fig. 6 gives an example of the cam and box type valve gear as used on a Durham locomotive constructed in 1822. The motion was conveyed through the cam rod to a bell crank lever which was connected to a double-ended lever fixed to the steam chest. Reversal of motion was obtained by the driver taking the pin out of point A and moving the cam rod to the other end of the lever and fixing pin in point B.[2] In 1880 the cam gear was replaced by an eccentric, a straight link providing the means of reversing engine as shown on trailing wheel of Fig. 6.

Before the steam engine was employed for marine or locomotive engines there was little need for reversing the direction of motion. The first locomotives were reversed in numerous ways, in some cases the engine was brought to rest and by means of a spanner the position of eccentric was changed by hand. The loose eccentric was a big improvement in the working of a locomotive. In this arrangement the eccentric was held in fore or back gear by stops on the driving shaft. Again, reversal was obtained by a fixed

---

1—Elemental Locomotion. A. Gordon, 1832.
2—Development of the Locomotive Engine. Angus Sinclair. 1907.

eccentric, the rod of which was capable of engaging at top or bottom of a rocking shaft. Later came the double eccentrics with gabs at the ends of eccentric rods to engage the valve spindle with either fore-going or back-going eccentric.

Fig. 7 shows the first known use of the eccentric valve gear on a locomotive, this being constructed by George Stephenson and used at Killingworth Collieries between 1814-15. The loose eccentric F caused the valve connections E to operate the slide valve of each cylinder. The eccentric was loose on the shaft, a lever fixed upon, and revolved with the driving axle, formed with a stud which entered and slid freely in a concentric groove cut in the eccentric sheave; the stud found its way to one end of the groove and determined the position of the eccentric for fore or back gear. The end of the eccentric rod had some play at its connection to the bell crank, whereby the valve was quickly opened and closed. Mr. Wood, then engineer of the Killingworth Engine Factory, invented the loose eccentric reversing motion.

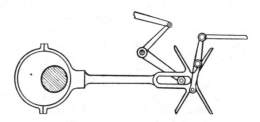

FIG. 8. CARMICHAEL. 1818.

Carmichael's, of Dundee, in 1818, introduced a form of gab motion, shown in Fig. 8, where a single fixed eccentric had a double forked eccentric rod, the forked end worked on a rocker arm having two actuating pins, one for fore and one for back gear. This was a very popular form of valve gear for marine, road carriage and locomotive use till the introduction of the double eccentrics twenty years later.

Bury, in 1828, used a form of drop gab motion, also with a single fixed eccentric, as shown in Fig. 9. It will be observed that each cylinder required two handles to perform the operation of reversal, the top lever to set the position of the valve for reversing and the lower horizontal lever to raise or lower the gab above or below the double-ended lever. From this and later figures it will be seen that two levers on the footplate were constantly in motion, and even now, over a century later, the term "rocking shaft" is a common phrase.

In 1826 Gurney obtained patents for improvements in locomotive engines. One form of steam carriage, Fig. 10, had a two cylinder engine working upon a crankshaft on the rear wheels of the vehicle. On the crankshaft were placed two eccentrics, J and K, to produce the motion for the slide valves placed above the cylinders, L being the steam pipes and D the exhaust pipes. By operating the lever N the engine could be reversed by moving the eccentric rod to engage above or below the double-ended rocking lever.

On this engine the steam was worked expansively by the following arrangement. A steel plate X was united to each eccentric and as eccentrics revolved X moved the connections of the expansion rod Y and the expansion lever, the cock at Z cutting off the steam to the steam chest at from one-half to two-thirds of piston stroke.[3] Gurney's expansion gear being probably the first of its kind on self-propelled vehicles.

Two years later, in 1828, Robert Stephenson built the locomotive " Lancashire Witch " for the Bolton and Leigh Railway, this engine also having an arrangement for working the steam expansively. On the rear axle was fixed a bevel wheel turning another wheel placed horizontally and driving a vertical shaft, this in turn operating a rotating plug valve. Through the medium of

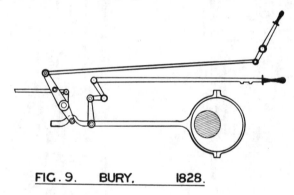

FIG. 9.   BURY.   1828.

two toothed quadrants the driver could either cut off steam at half-stroke or full-stroke.[4] However, as the engine was supposed to use the half-stroke cut-off when steam was low when commencing duty, this apparatus can hardly be classed as for expansive working.

What may be termed the father of radial valve gears was that used by Melling, of the Liverpool and Manchester Railway, in 1830. Fig. 11 shows that the eccentric was replaced by a stud midway on the connecting rod on which was attached a slotted lever, by this connection a slip on the connecting rods motion reduced the travel of the other end of the lever sufficiently to give the required movement for working the slide valve. The centre of the stud on the connecting rod described an ellipse, however, the motion imparted to the valve was at its slowest when it should have been greatest, at the points of admission and cut-off. To overcome this defect Melling amended his gear so that each slide valve received

---

3—Elemental Locomotion. A. Gordon, 1832.
4—A Century of Locomotive Building. Warren.

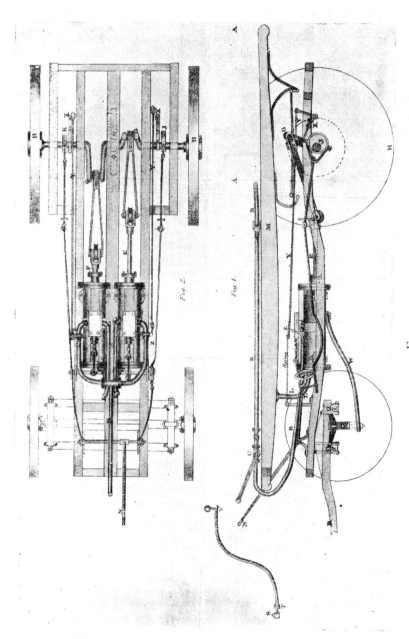

Fig. 10.

Gurney's Expansion Valve Gear, 1826.

its motion from the connecting rod of the opposite cylinder, and this is supposed to have improved the working of the gear.

Bourne, in 1834[5], experimented with a form of link motion and also worked on the idea of giving the slide valve an adjustable lap. He later came to the conclusion that the desired end could be attained by altering the travel of the valve, whereby the fixed lap would be relatively greater or less. It finally occurred to him that a double-ended lever moved by an eccentric and with a slot along its face contrived to impart the motion to that part of the link or lever from which the eccentric rod proceeded to move the valve, would be a satisfactory arrangement. Bourne admits that a double-ended lever had been used before this date for reversing engines, and on examining an old Napier steam boat at Greenock, he obtained the idea of the reversing gear he adopted. All previous gears had the sole function of the double-ended lever to reverse the engine and were not capable of altering the travel of the valve. Fig. 12 shows Bourne's gear, and its form preceded a somewhat similar arrangement (Fig. 47) by Florian Angeles, of Paris, in 1843. This latter gear, however, used a return crank in place of an eccentric.

Hawthorn introduced in 1838 his slotted link and connecting rod motion. This was somewhat similar to Melling's gear, the connecting rod having a stud on it, and the working lever slotted where attached to the stud, Fig. 13, the lead of the valve was regulated by the inclination of the straight link, which was adjustable.

Slide valves at this period had little or no outside lap, the admission of steam to the cylinders occurred for nearly the whole of the piston stroke. Locomotive engineers were put to severe test to reduce the fuel consumption, and about 1838 slide valves were made with from 1-16th in. to ⅜ in. outside lap, and two years later an outside lap of one inch was adopted on some engines, reducing the fuel consumption considerably.

Mr. John Gray, of Newcastle, in 1839, applied a form of expansion gear to the locomotive " Cyclops " on the Liverpool and Manchester Railway, for which an economy of 12 per cent. in fuel was claimed. Fig. 14 represents this so-called " horse leg motion." The pin of the eccentric rod worked in a segmental lever curved to the radius of the rod, the upper end of which was linked to the valve spindle. Double eccentrics were employed, the slotted lever being concentric with the foregoing eccentric rod at the beginning of the stroke. The eccentric rod could be raised or lowered in the slope of the lever to any required distance from the fulcrum, thus regulating the travel of the valve while the lead remained constant. The reversing mechanism consisted of a wrought iron frame which slid horizontally on two fixed pivots and carried rollers working in grooved levers linked to the eccentric rods. The eccentric rods could thereby be placed in and out of gear by the reversing rod. Gray's

---

5—Steam, Gas and Air Engines. Bourne, 1878.

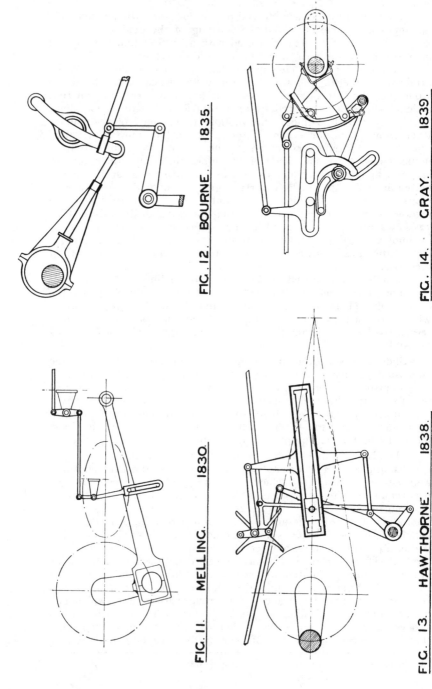

FIG. 11. MELLING. 1830.

FIG. 12. BOURNE. 1835.

FIG. 13. HAWTHORNE. 1838.

FIG. 14. GRAY. 1839.

valve gear was later used on the York and North Midland Railway in 1840, and on the Hull and Selby Railway in 1842.

Three views of a locomotive of the period 1830-38 using drop hook motion with one loose eccentric for each cylinder is shown in Fig. 15, this giving a typical example of the most common form of valve gearing before the introduction of double eccentrics. To place slide valve in position for the required direction of motion the driver moved the lever Z and the vertical lever Y lifted the gab end of the eccentric rod till the rod engaged with the slide valve rocking lever. The two levers Z on the footplate, one for each valve, were constantly moving backward and forward while the engine was in motion. As engine speeds increased the onerous duties with such a valve gear became difficult for enginemen to carry out, and in 1834 two eccentrics were introduced, one for fore and one for back gear for each cylinder. William T. James, of New York, claimed to have used double eccentrics in 1832[6]; Forresters, of Liverpool, built the locomotive " Swiftsure " for the Liverpool and Manchester Railway and also the " Vauxhall " for the Dublin and Kingston Railway in 1834, the valve gear in both engines having four fixed eccentrics and four separate gab-ended eccentric rods[7], this arrangement displacing the constant moving lever Z (Fig. 15) on the engine footplate and allowing enginemen more freedom to operate the locomotive. R. and W. Hawthorn built in 1835 the locomotive " Comet " for the Newcastle and Carlisle Railway in which four fixed eccentrics were employed. According to D. K. Clark the first application of four eccentrics to locomotives was made by Hawthorn's, of Newcastle, in 1837, and he adds that an unknown mechanic in Newcastle had, previous to this date, constructed a model valve gear for locomotives, using four eccentrics.[8]

Fig. 15a shows Bury's double eccentric gab motion as fitted to his 2-2-2 type locomotive of 1837.[9]

Fig. 16 illustrates the double eccentric type of motion by Stephenson's in 1838, each eccentric rod having a vee shaped open end, usually diametrically opposed, but in arrangement shown both eccentric rods face upward; the reversing arrangement to lift whichever eccentric was required to engage the rocker arm can be followed from diagram.

In 1834 Forrester used a valve motion in which one eccentric acted vertically (Fig. 36) the gear being provided with two diverging rods furnished with gabs, and these were placed in gear on one side or the other by a double spanner or held out of gear by a double-ended pall hung between them and operated from the footplate. Fig. 17 shows a simple form of gab motion, not unlike Bury's, this being due to Pauwel in France. Other gab motions of this period and of somewhat similar form were Sharp's, Roberts', Hawthorn's, Buddicorn's and Jackson's. Finally, in 1840, Stephenson improved his

---

6—Development of the Locomotive Engine, page 119. A. Sinclair.
7—The British Steam Locomotive, pages 29-30. E. L. Ahrons, 1927.
8—Railway Machinery. 1855. D. K. Clark.
9—Journal of Arts and Sciences. 1858.

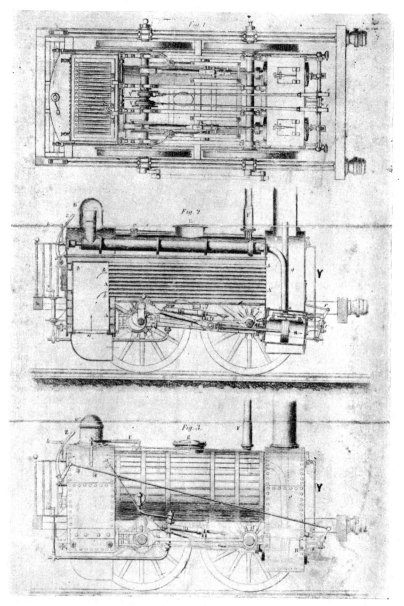

FIG. 15.
LOCOMOTIVE WITH DROP-HOOK MOTION.
PERIOD: 1830-1838.

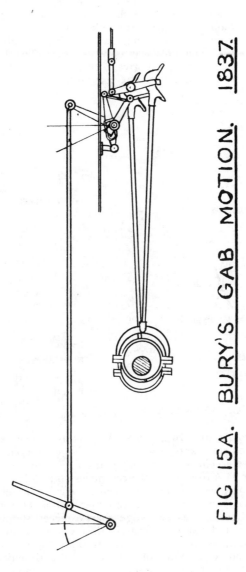

FIG. 15A. BURY'S GAB MOTION. 1837.

valve gear by placing the gabs on the valve spindle, as in Fig. 18, and a straight link joined the ends of the fore-going and back-going eccentric rods, this being the forerunner of the now common link motion, although the idea of the link was proposed in various forms before the invention of Stephenson's link motion.

The form of link used by Rodgers in the U.S.A. about 1844 is given by Fig. 19. The link was only capable of working in full fore or back gear. Also Fig. 20 gives a form of gab motion common in America during the late thirties of last century.

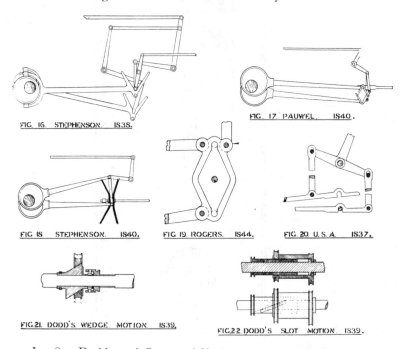

FIG. 16. STEPHENSON. 1838.
FIG. 17. PAUWEL. 1840.
FIG. 18. STEPHENSON. 1840.
FIG. 19. ROGERS. 1844.
FIG. 20. U.S.A. 1837.
FIG. 21. DODD'S WEDGE MOTION. 1839.
FIG. 22. DODD'S SLOT MOTION. 1839.

In 1839 Dodds and Owen, of York, patented their first form of reversing motion now known as Dodds' wedge motion, Fig. 21.[10] In this gear the eccentrics were mounted on the crankshaft between guide pieces or ears to prevent the eccentrics sliding laterally on the shaft but allowed them to move between the ears transversely. The parts that moved the eccentrics more or less from the centre of the shaft were inclined planes or wedge shaped pieces sliding laterally upon the shaft and passing through the square openings in the eccentrics so that, as they were forced through the eccentrics, they changed the relative positions of the eccentrics and shaft. Fig. 22 gives another form of Dodds' motion. This was mounted on the crankshaft and acted upon the eccentrics so as to change their position. An inner tube embraced the shaft and mounted so as to

---

10—Link Motion and Expansion Gear. N. P. Burgh, 1870.

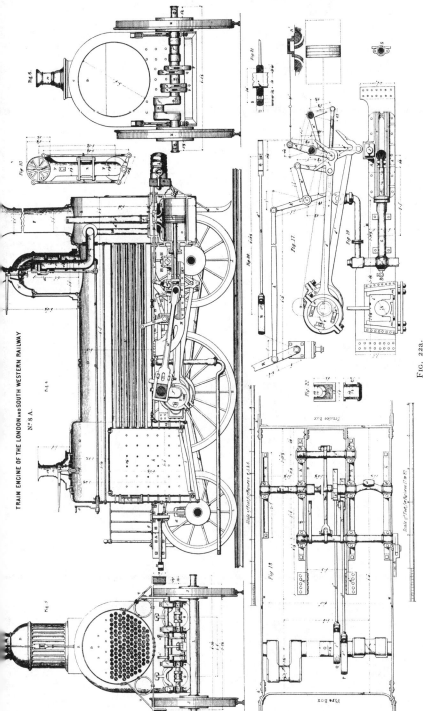

Fig. 22a.

GAB MOTION, FAIRBAIRN, MANCHESTER.

slide along it and at the same time partly round it, like the motion of a screw thread. In the tube was formed a curved slot and upon the inclination of the curve of this slot depended the movement of the eccentric. On the shaft was fixed a stud which projected into the groove. When the tube was moved along the shaft the stud caused it to move round the shaft. The eccentrics were fixed upon another outer tube embracing the inner tube, but was prevented from moving along the shaft by means of a collar fixed on the shaft coupling piece. The outer tube was fixed to the inner tube by means of a feather and a slot capable of one sliding along the other, so that, whenever the

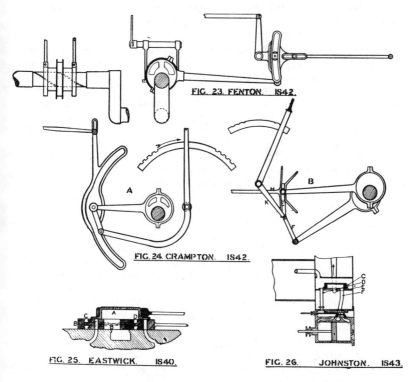

FIG. 23. FENTON. 1842.

FIG. 24. CRAMPTON. 1842.

FIG. 25. EASTWICK. 1840.

FIG. 26. JOHNSTON. 1843.

inner tube was moved along the shaft it would be drawn out of the interior of the outer tube and in consequence of the pin on the shaft being in connection with the curved groove the inner tube was made to move round the shaft, and the inner tube by means of its connection with the feather and slot caused the outer tube to move with it round the shaft, the eccentric was made to move into another position changing its place on the shaft.

Originally Dodds' wedge motion was not intended for variable expansion, but with a slight adjustment expensive working was obtained.[11] The Dodds' wedge motion was amended and used by

---

11—See " Locomotive Engineering." Colburn.

Dubs and Co., of Glasgow, between 1842-52, and it was also used as late as 1872 on the North Staffordshire Railway. Fig. 22a shows a form of gab motion as fitted to early L. & S.W. Railway locomotives and built by W. Fairburn, of Manchester.

Fenton, in 1842, introduced a combined sliding eccentric and link motion, Fig. 23. The eccentrics were fixed on the shaft by a spiral feather which adjusted themselves to the angle for the required cut-off. One eccentric was used for each cylinder and the eccentric rod was attached to a link vibrating on a fixed centre. The link was slotted to receive the end of the radius rod attached to the valve spindle. The position of the radius rod in the link regulated the degree of expansion and the lead was kept constant by a simultaneous adjustment of the eccentric.

Crampton's valve motion of 1842 was a development of Carmichael's gear, one eccentric being used for each cylinder. Fig. 24a shows the eccentric out of gear, the end of the eccentric rod working in the widest part of the link but imparting no motion to it. When the reversing lever was moved in the direction of the arrow to the first notch in the quadrant the end of the eccentric rod was guided upward into the narrow part of the groove until the eccentric rod formed an angle of 20 degrees with the centre link and in this position the eccentric rod moved the link so as to cut off steam in the cylinder at 85 per cent. of piston stroke. By moving the quadrant lever to the last notch the end of the eccentric rod was moved to the top of the grooved link and formed an angle of about 60 degrees with the centre link, the valve then cut off steam at about 30 per cent. of the piston stroke. Moving the reversing lever in the opposite direction to that of the arrow the eccentric rod engaged with the lower part of the link and the action of the slide valve was reversed. Fig. 24b gives a double eccentric valve gear also by Crampton, the eccentric rods were joined by two slings, E and F, meeting the common arm of the reversing lever K, the valve spindle H having double gab ends and grooved from gab to gab, the curve of the groove being taken from the driving axle. The position of the eccentric rod in the grooved slot controlled the cut-off of steam to the cylinders and when reversing lever was in centre of quadrant neither eccentric rod was engaged with valve spindle and no movement was imparted to the valve. There is no record of either of these two gears being used in actual operation.

A novel form of reversing " gear " was invented by Eastwick, of Philadelphia, in 1840, and adopted on the first steam locomotives for the Nicolai Railway from St. Petersburg to Moscow. In Fig. 25a, is a slide valve and B a sliding, movable seat. When this block or seat was fixed in position shown the lap edge of the valve A admitted steam to the front of the cylinder. By moving the sliding block B backward till port C coincided with the steam port on the cylinder the steam admitted to the front edge of the slide valve was conducted by cross ports in the sliding block to the back of the cylinder and steam admitted to the rear of the sliding block at

D was conducted to the front of cylinder thereby reversing the motion of the engine.[12]

A somewhat similar arrangement was proposed by William Johnston, of Preston, in 1842,[13] by which the exhaust cavity in the slide valve became the steam chest (inside admission) as a means of reversing the locomotive. Fig. 26 is a rough sketch of this proposal. Steam from the dome was conveyed to a valve box in the smokebox. The three pipes, D, E and F opened into this box,

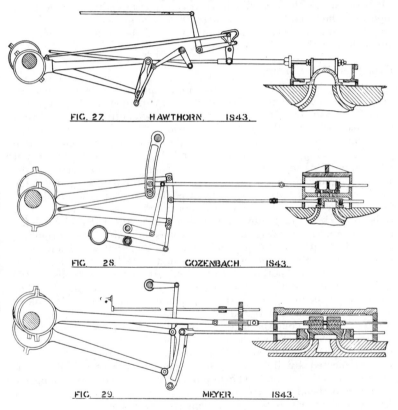

FIG. 27.   HAWTHORN.   1843.

FIG. 28.   COZENBACH   1843.

FIG. 29.   MEYER.   1843.

two of which, when the engine was in motion, were always covered by the large slide valve C. The uncovered passage admitted steam to the steam chest of the cylinders. In the position shown steam entered the pipe D into the steam chest and from there into the cylinder behind the piston as indicated by the arrow, the exhaust at the same time escaping by the exhaust pipe F into the hollow of the slide C, thence to the pipe E into the blast pipe. By shifting

---

12—See " Locomotive Magazine," Volume XXI, page 179.   Volume XLI, page 299.
13—Glasgow " Practical Mechanic," January, 1842.

the large slide valve C till it covered the passages D and E leaving F open to the valve box, the action of the steam was reversed, the steam pipe D became the exhaust pipe and the exhaust pipe F the steam pipe. Provision was made for keeping the slide valve down on its seat when the exhaust cavity of the engine slide valve became the steam chest. When steam was cut off and engine coasting, the atmosphere was given free access to the cylinders by making the large slide valve C of sufficient length to cover all the three passages, when the atmosphere would have a free passage through the blast pipe into the cylinders. Of course this arrangement could only be possible with an eccentric at right angles to the crank and with no lap or lead with the slide valve.

Expansion Valve Gears.

Before dealing with the link motion proper, a glance may be taken at a few arrangements of valve gears contrived to use the steam expansively in the cylinders of the locomotive with, and in a few cases without, the use of expansion valves in addition to the slide valve. Hawthorn's, of Newcastle, improved on their original valve gear (Fig. 13), and this amended gear was the first to cut off steam by a separate valve to the slide valve, on locomotives. The arrangement shown in Fig. 27 was adopted in 1843, the slide valve was placed horizontally and was worked from the crankshaft by two eccentrics. The link was contrived to allow the eccentric rods to work without moving the slide valve, the valve closing the ports in mid-gear position. The expansion valve was fitted to, and worked upon, the same surface as the slide valve. There were projections on the inside of the expansion slide frame so that when the frame was in motion it overlapped alternately the ends of the slide valve according to the degree of expansion required. The valve rod was attached to the expansion frame and made hollow in order that the valve rod of the main slide valve would pass through it, or it could be made solid and worked through a separate stuffing. A short weigh bar supported the two levers that worked the expansion frame, the lower end of one of the levers was connected by links and the expansion eccentric rod to the expansion valve frame. The upper end of the lever was connected by the expansion eccentric rod to the back-going eccentric. There was a slot in which the pin at end of expansion rod moved and also gearing by which the driver could vary the position of this pin and cause the stroke of the expansion slide frame to correspond with the amount of expansion required. A dial was placed on the footplate to indicate the amount of expansion at any part of the stroke, as in modern practice. When expansion was not required, as at starting, the pin at end of expansion eccentric rod was raised into a loop and at the same time the inclined surfaces at end of the lever pressed against the pins, which kept the expansion slide frame secured and at rest when out of gear.[14]

14—See Link Motion and Expansion Gear. N. P. Burgh, 1870.

J. Bourne, in his patent of 1838, claims to have preceded Bodmer and Meyer expansion valves of 1841 and 1842 respectively.[15] Bourne's expansion valve was in the form of two plates connected with a spindle which had a right and left-handed screw, and by turning the spindle the plates could be advanced or retarded, thereby varying the rate of expansion. Auxiliary expansion valves were commonly used on the Continent and U.S.A. till the middle fifties and in a few cases till 1873, even in the twenties of the present century examples of separate valves for expansive working on locomotives have been patented.

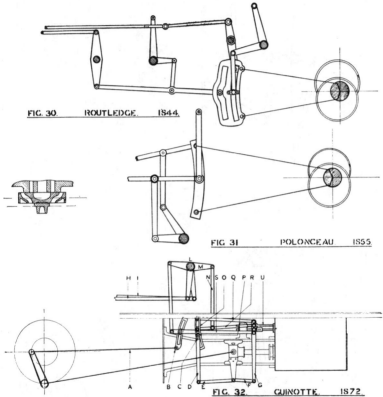

FIG. 30. ROUTLEDGE. 1844.

FIG. 31. POLONCEAU 1855.

FIG. 32. GUINOTTE. 1872.

In Germany, Gozenbach patented, in 1843, a form of expansion gear used with double eccentric gab motion, and this was later improved, as shown in Fig. 28. The two links were of the box and slotted type respectively and performed separate duties. The box link was in communication with the main slide valve, this valve being used to effect the forward or backward working of the locomotive, or for bringing engine to rest. The intermediate

15—Steam, Air and Gas Engines. J. Bourne, page 13.

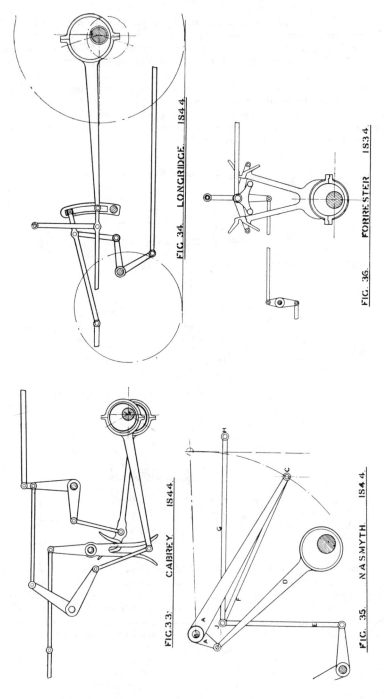

positions of the link were not employed to govern the movement of the valve to produce variable expansion, this being attained by the supply or expansion valve. The expansion valve eccentric rod was fastened to its end to the slotted link and was connected as shown, to the back-going eccentric. The link was suspended from a fixed point above it and received an oscillating motion, the end of the radius rod was connected to the spindle of the expansion valve and transmitted the movement of the slotted link to this spindle. The nearer the die block was to the top of the link the shorter the stroke of the expansion valve. The expansion valve moved in a separate steam chest, and worked on a seating having two port openings connecting the upper steam chest from the main steam chest. When the ports in the expansion valve coincided with the ports in the expansion valve seating the steam entered the steam chest of the main slide valve, expansion beginning as soon as the expansion valve shut these openings. The slots in the expansion valve were made a little larger than the openings in the seating. This gear was also used in the U.S.A. under the name of Cuyahoga cut-off gear. In 1844 Nasmyth and Caskell, of Manchester, introduced a cut-off gear similar to above in which the cut-off gear link was worked from the crosshead of the contrary cylinders by means of suitable reduction levers, the expansion valve working on the back of the perforated main slide valve.[16]

Thomas Rodger, of New Jersey, in 1843, designed an expansion gear with a separate chamber for the expansion valve; however, this gear was too complicated for general use and after building his locomotives with a link capable of working only in full fore and full back gear (Fig. 19), Rodger, in 1849, reverted to a link motion of the Gooch type, but even after this date we find him using a form of Gozenbach's expansion gear in conjunction with Stephenson's link motion.

In 1845 Baldwin's introduced an expansion gear also after the Gozenbach pattern, in which the upper part of the valve was operated by a separate eccentric, this eccentric being thrown out of gear when not required. The single expansion valve eccentric was usually arranged to cut off steam at half-stroke and was mainly used on passenger locomotives. Eight years later Baldwin's adopted a variable cut-off gear consisting of two valves, the upper sliding on the lower and worked by an eccentric and rocking shaft in usual manner. The upper arm of the rocker arm was curved so as to form a radius arm on which a sliding block formed the end of the upper valve rod. This could be adjusted at varying distances from the axis producing various travel of the expansion valve.[17]

John Stevens, in the U.S.A., used a form of expansion gear in 1849 in which the main slide valve was operated by two eccentrics and the expansion valve by a return crank on the outside crank pin, the motion being transmitted to the expansion valve through a rocker arm.

16—Glasgow " Practical Mechanic," 1846.
17—Development of the Locomotive Engine. A. Sinclair, 1907.

On the Continent expansion valve gears were usually fitted to locomotives, these being of the German Gozenbach double link type or of the French Meyer three eccentric type. The latter gear is shown in Fig. 29 in an improved pattern, the original gear, being like Gozenbach's, fitted with gab-ended eccentric rods. In the Meyer gear three eccentrics were used for each cylinder, the main slide valve receiving its motion from the usual form of link motion, the link only being used for reversing purposes. On the back of the flat main slide valve were two ports to allow the steam to enter the main cylinder ports, when the expansion valve did not cover the corresponding slide valve ports. The expansion valve consisted of two plates with right and left-hand screw threads and were connected with the expansion valve spindle provided with screw threads similar to the expansion valve. By turning the valve spindle by a handle on the footplate the two plates would either move further apart for expansive working or approach each other to prevent expansive working of steam in the cylinder.

Henry Routledge, of the Grand Junction Railway, used a form of valve gear, Fig. 30, in 1844-45, in which the cut-off valve worked on a separate stationary face above the slide valve, the cut-off valve being simply a plate with a lap of $\frac{7}{8}$ in. A double slotted link was used, the front slot of the link controlled the expansive working of the steam while the slot in the link next the crankshaft was used for either full fore or back gear.

The Paris and Orleans Railway, in 1855, used a novel addition to a gear of the Gooch type, as shown by Fig. 31. No improvement was made on the admission events but a constant exhaust was attained by using two die blocks working to a parallel radius in a double slotted link. The control was operated by two separate levers one of which was only concerned with reversing the engine while the other varied the cut-off required by controlling the movement of the small expansion valve working on back of slide valve as shown. This gear was called the Polonceau link motion.

Fig. 32 gives an example of an expansion valve gear of the Walschaerts type used on the Belgian Grand Central Railway in 1872. This arrangement, due to Guinotte, was designed for outside admission.[18] The eccentric rod A worked the main expansion link B, the lead being obtained in the usual manner through the combination lever G. The main valve was only concerned with reversing the engine and the exhaust, the die block on the main link having only two running positions of full fore and back gear, the exhaust being constant for all degrees of expansion. The expansion valve gave a cut-off from 10 to 70 per cent., this being provided by the expansion link O placed in front of the main link. The upper end of the expansion link O was connected by a rod Q to an extension of the upper end of the combination lever G, while link oscillated about a point P, the bearing so given being the upper end of the rocking lever D, whose fulcrum was in bearing C,

18—" The Locomotive," Vol. 44.

carried in an extension of the main link bracket. The lower and longer arm of the lever D was connected by rod E to the engine crosshead, and the expansion link O had a compound movement which gave rapid acceleration of the expansion valve from the spindle U and the expansion radius rod R. The radius rod R could be moved by the lifting rod N and a bell crank lever M mounted on a hollow weigh bar shaft L, through which passed another weigh bar shaft operating the reversing radius rod S. The reversing rod H worked in a quadrant, having three positions only, for forward, mid and back gear. The expansion rod I being actuated by a screw to set the cut-off as required.

Cabrey, of York, in 1844, introduced a valve gear similar in its effects to the link motion; this he termed the " sliding fork " gear. In Fig. 33, the slide valve was connected by the valve rod to the upper end of the slotted rocking lever which worked upon a fixed centre, its lower end terminating in a forked end. This rocking lever was worked by the forward eccentric rod, this eccentric rod being provided with a pin fitting the slot in the forked lever. The end of the eccentric rod was suspended from a lever keyed on the weigh bar shaft which also carried a second lever attached to a link keyed on the second weigh bar shaft. The latter carried a similar lever provided with a supporting link for the backward eccentric rod. The back-going eccentric rod carried a fork on its extremity arranged to gear when required with a pin on a second lever on the rocking shaft so that for backward motion the engine worked without expansion. The lead of this valve gear was lost when the pin of the foregoing eccentric rod approached the fulcrum of the sliding fork. Cabrey's gear was used on Belgian and Dutch railways.

A valve gear after the pattern of Fenton (Fig. 23) was used by Longridge in 1844. From Fig. 34 it will be seen that a link was made use of to vary the travel of the cut-off valve while the slide valve preserved its full travel. The eccentric working the slide valve served the double purpose of a back and forward motion, being made movable round the crankshaft so as to take up either a back or forward position. This was accomplished by the driver working a spiral groove, cut in the inside of the eccentric, in guides or in a groove cut longitudinally along the axle, to revolve the eccentric on the axle to the position required. The cut-off valve was a single plate perforated in the centre and worked on back of slide valve, the curved link, operating the cut-off valve, received its motion from another eccentric, the pin in the link being raised or lowered to suit the required travel of the cut-off valve.

Shortly after the introduction of the curved link James Nasmyth proposed an arrangement of levers as a substitute for the link. In Fig 35 A is a double-ended cranked lever vibrating at the centre B, the length of this lever from its centre B to the joint C connecting the valve spindle was the same length as the radius of the curved link would be if similarly applied. D is the connecting rod by which the lever A is worked either from the piston crosshead or from an eccentric. The expansive working of the steam was effected by raising or lowering the rod E joined to the link F carried by the lever A at C. The connecting link G was also joined to the same pin and to the cut-off valve spindle at H. It will be seen that when the motion was communicated to the lever A by the rod D the end C would move in a curve proportional to its length, and as the link F was carried by it, the end J would also move in another curve proportional to its distance from the centre of motion B. If, therefore, the end J was moved nearer to or further from this centre, the stroke of the cut-off valve would be similarly diminished or increased. The idea embodied in the above was later made use of by other inventors of valve gearing.[19]

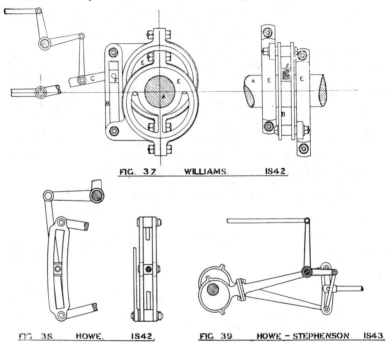

FIG. 37. WILLIAMS. 1842.

FIG. 38. HOWE. 1842.

FIG. 39. HOWE – STEPHENSON. 1843.

Link Motion.

American writers claim that the link motion with double eccentrics was the invention of William T. James, of New York, in 1832. On this side of the Atlantic the credit of the invention is generally attributed to Mr. Williams, a " gentleman apprentice " of Robert Stephenson's, or to William Howe, a pattern-maker with the same firm. The foregoing controversy is a century old; according to U.S.A. historians W. T. James employed the " link motion " on road and rail vehicles between 1828-32, and after the failure of these engines James sent to Forrester's, of Liverpool, drawings of

---

19—Glasgow " Practical Mechanic," 1846.

THE EVOLUTION OF LOCOMOTIVE VALVE GEARS   25

the double eccentrics, and there is a " chance " that the link was also shown. Again, when James tried his engine on the Baltimore and Ohio Railway in 1832, foreign visitors would probably notice the form of valve motion used and keep this for further reference.[20] Irrespective of the foregoing, the link motion was not used again in America till Baldwin's built a few engines with link motion in 1845, but it was not till 1854 that this firm made general use of the Stephenson link, and although Rodgers used a form of link motion in 1849, separate expansion valve gear was the general valve gear in U.S.A. till 1854, when the undisputed value of Stephenson's link motion was too evident to be neglected.

Regarding the Williams-Howe controversy, Howe has stated[21] that in the mid-summer of 1842 a species of link motion, Fig. 37, was suggested by Mr. Williams. In the figure A is the crankshaft, B the proposed link, C the eccentric rod, E are the eccentrics, and F the die block for reversing the motion of the slide block. This suggestion could be of no practical value as one eccentric would displace the other while in motion. Several persons at the works of Robert Stephenson's saw this drawing and no one brought the gear into practical use till August, 1842, when Howe made a sketch and rough wooden model of the link motion, Fig. 38. Howe showed this model to Mr. Hutchieson, the manager of the works, who at once saw the worth of its application. The model was then sent to Robert Stephenson, who was in London, and he approved of the gear immediately. At this time two locomotives were being constructed by Stephenson's for the North Midland Railway, and these were to be fitted with Dodd's wedge motion. One of the locomotives, N.M. Rly., No. 358 (Builders No. 70) had the wedge motion fitted, but No. 359 (Builders No. 71) was fitted with Howe's link motion. The valves of this engine had ½ in. outside lap and 1-16th in. inside lap, the throw of the eccentric was 1½ in., steam port opening 1 in. and length of eccentric rod 5 ft.

TABLE I.

STATEMENT OF THE WORKING OF HOWE'S LINK MOTION LOCOMOTIVE NO. 71 ON THE NORTH MIDLAND RAILWAY,
CONSTRUCTED IN 1842.

Notch.	Valve Travel.	Steam Port Open.	Supply Lead.	Cut-off at Stroke of Piston.	Port Open.	Length of Expansion.	Length of Exhaustion.	Exhaustion Lead.
1	$2\frac{3}{4}$	$\frac{7}{8}$	$\frac{1}{8}$	$17\frac{1}{8}$	$19\frac{1}{8}$	2	$\frac{7}{8}$	$\frac{9}{16}$
2	$2\frac{1}{2}$	$\frac{3}{4}$	$\frac{1}{8}$	$16\frac{1}{4}$	$18\frac{3}{4}$	$2\frac{1}{2}$	$1\frac{1}{4}$	$\frac{9}{16}$
3	$2\frac{1}{4}$	$\frac{5}{8}$	$\frac{1}{8}$	$14\frac{7}{8}$	$18\frac{1}{8}$	$3\frac{1}{4}$	$1\frac{7}{8}$	$\frac{9}{16}$
4	2	$\frac{1}{2}$	$\frac{1}{8}$	13	$17\frac{1}{8}$	$4\frac{1}{8}$	$2\frac{7}{8}$	$\frac{9}{16}$
5	$1\frac{3}{4}$	$\frac{3}{8}$	$\frac{1}{8}$	$10\frac{3}{8}$	$15\frac{3}{4}$	$5\frac{3}{8}$	$4\frac{1}{4}$	$\frac{9}{16}$
6	$1\frac{1}{2}$	$\frac{1}{4}$	$\frac{1}{8}$	$7\frac{3}{4}$	14	$6\frac{5}{8}$	6	$\frac{9}{16}$
7	$1\frac{1}{2}$	$\frac{1}{4}$	$\frac{1}{8}$	$4\frac{1}{4}$	$11\frac{1}{2}$	$7\frac{1}{4}$	$8\frac{1}{2}$	$\frac{9}{16}$

Cylinders 14 in. dia. × 20 in. stroke.
Outside lap ½ in.   Inside lap 1/16 in.

20—Angus Sinclair.
21—History of Link Motion. N. P. Burgh, 1870.

TABLE II.

Statement of the Working of Dodd's Wedge Motion Locomotive No. 70 on the North Midland Railway, Constructed in 1842.

Notch.	Valve Travel.	Steam Port Open.	Supply Lead.	Cut-off at Stroke of Piston	Port Open.	Length of Expansion.	Length of Exhaustion.	Exhaustion Lead.
1	$3\frac{3}{4}$	$\frac{7}{8}$	$\frac{1}{8}$	$14\frac{1}{2}$	$18\frac{1}{8}$	$3\frac{5}{8}$	$1\frac{7}{8}$	1
2	$3\frac{11}{16}$	$\frac{13}{16}$	$\frac{1}{8}$	$14\frac{1}{8}$	$18\frac{1}{8}$	4	$1\frac{7}{8}$	1
3	$3\frac{1}{2}$	$\frac{3}{4}$	$\frac{1}{8}$	13	$17\frac{7}{8}$	$4\frac{7}{8}$	$2\frac{1}{8}$	1
4	3	$\frac{1}{2}$	$\frac{1}{8}$	11	$16\frac{7}{8}$	$5\frac{7}{8}$	$3\frac{1}{8}$	1
5	$2\frac{9}{16}$	$\frac{1}{4}$	$\frac{1}{8}$	$7\frac{1}{2}$	$15\frac{1}{2}$	8	$4\frac{1}{2}$	1
6	$2\frac{1}{4}$	$\frac{1}{8}$	$\frac{1}{8}$	$3\frac{1}{4}$	$12\frac{1}{4}$	9	$7\frac{3}{4}$	1
7	$2\frac{3}{16}$	$\frac{1}{8}$	$\frac{1}{8}$	1	$8\frac{7}{8}$	$7\frac{7}{8}$	$11\frac{1}{8}$	1

Cylinders 14 in. dia. x 20 in. stroke.   Steam ports 10 in. x 1¼ in.
Outside lap 1 in.   Inside lap 1/16 in.

Tables I and II give the valve events with link motion for engine No. 71 and for No. 70 with Dodd's wedge motion. These tables dispose of the statement often heard, that when Howe joined the ends of the eccentric rods by the curved link the working of the link as an expansive medium was not then understood nor anticipated. Fig. 39 shows the first application of the Howe-Stephenson link motion in September, 1842. William Howe was granted a present of 20 guineas by his firm, and in 1846 he was appointed engineer of the Stephenson Collieries at Chesterfield, and a year later he applied his link motion to a stationary winding engine.

The Glasgow "Practical Mechanic," during 1846, published a series of articles on "Expansion Valves and Gears," and in the April number they stated:—

"From the comparatively complicated specimens of mechanisms we now turn to one which, while it is practically equally effective in obtaining the full expansive effect from the steam it is also much cheaper and less liable to get out of order. The arrangement referred to is the 'Link Motion,' invented by Mr. Williams and applied by him as a reversing and expansive gear to locomotive engines."

Then followed a description of Stephenson's link motion as in Fig. 39. In this magazine one month later the same writer stated:

"Since writing the foregoing description we have received a communication from Mr. William Howe, of Newcastle, relative to the ownership of the invention of 'Link Motion.' In our description of the expansion gear referred to, we ascribed the invention to Mr. Williams. This is contradicted by Mr. Howe, who himself claims the invention of the 'Link' as originally applied by Mr. Robert Stephenson to locomotive engines. The original idea, Mr. Howe concedes, to Mr. Williams, and he has forwarded to us a drawing[22] of the later

---
22—Fig. 37.

THE EVOLUTION OF LOCOMOTIVE VALVE GEARS

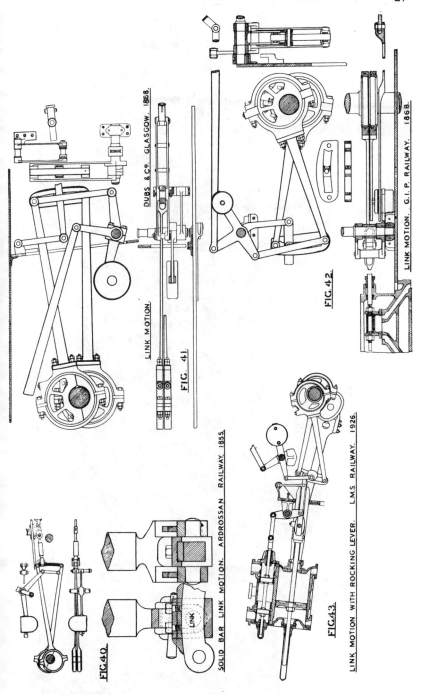

FIG. 40.
FIG. 41. LINK MOTION. DUBS & CO. GLASGOW. 1868.
FIG. 42. LINK MOTION. G.I.P. RAILWAY. 1868.
FIG. 43. LINK MOTION WITH ROCKING LEVER. L.M.S. RAILWAY. 1926.
SOLID BAR LINK MOTION. ARDROSSAN RAILWAY. 1855.

Fig. 44.

gentleman's scheme, which was never carried into effect. Mr. Howe informs us that this scheme was never put into practice, but leaves us to presume that it formed the ground for the invention of link motion."

The foregoing was written only three and a half years after the link motion made its appearance, and its conclusion may be well taken as all that can be said on the Williams-Howe controversy.

The Stephenson link motion has been almost universally applied with different modifications up to the present date despite scores of other valve motions patented or proposed, each claiming some advantage over the link motion. However, during the last quarter of a century Walschaerts radial valve gear has been more popular, especially with outside cylinder engines. One modification of the link motion is shown in Fig. 40, this being the solid bar link motion fitted to the locomotive " King Coil " in 1851 on the Ardrossan Railway, Ayrshire. The gear was suspended and reversed by a link attached to the back-going eccentric rod. Fig. 41 gives a design by Dubs and Co., of Glasgow, in 1868, a form that was standard on the G. & S.W. Railway in the seventies of last century ; here the link was suspended from the reversing shaft placed below the gearing, the intermediate valve spindle or valve rod was suspended from a link attached to a bracket on the motion plate above the gearing. This gear was commonly called the " swinging link " motion. The link motion in Fig. 42 had the valve rod suspended from a link attached to the reversing shaft placed above the motion, the gearing being suspended by a link with a pin common to the back-going eccentric rod. A modern form of link motion with rocking shaft and actuating inside admission piston valves is shown by Fig. 43, the link is suspended from the centre by the reversing arm link, and the intermediate valve rod is suspended from a link attached to a bracket on the motion plate.

The use of return cranks projecting from the main crank or coupling rod pins to operate the link motion has been proposed by Crampton in 1845 (Patent No. 10854), Crosby in 1860, and later by Anderson in the U.S.A. The practice of having the link motion outside the engine frame was common on the Continent and in a lesser degree in this country during the middle of last century, an example being given by Fig. 44, this being Crampton's locomotive " Kinnaird " on the Dundee and Perth Railway in 1846. Fig. 45 shows the Stephenson link motion as fitted to a Crampton intermediate driving shaft locomotive patented in 1849 and built in 1851 by R. Stephenson and Company for the S.E. Rly.; and by R. B. Longridge and Company for the G.N. Rly.

Gooch Link Motion.

The " stationary link motion " invented by Daniel Gooch in 1843 is shown in plan and elevation by Fig. 46 in which the eccentrics and eccentric rods were similar to Stephenson's gear; in the latter gear the convex side of the link was turned towards the shaft, while in the Gooch gear the convex side of the link was turned from the shaft. The Gooch link was hung from the centre by a swing link or radius bar on each side, so that the travel of the centre of the link, due to the motion of the eccentrics, was always on an arc described by the suspension links. The die block in the link was connected to the valve or radius rod, and this was raised and lowered by levers from the reversing quadrant. In this case the shifting of the die block instead of the link controlled the direction of motion of the engine and the expansive working of steam in the cylinders. The Gooch motion was widely used in this country and it became very popular throughout Europe.

Walschaerts Valve Gear.

On the Continent, in the forties of last century there appeared a form of valve gear that was to become, a century later, almost universal in locomotive operation, this motion being due to Egide Walschaerts, a Belgian mechanic, in 1844, and also to von Waldegg, in Germany, in 1849.

In France, Florian Angele patented in 1843[23] an original form of valve gear dated November, 1842. From Fig. 47 it will be seen that the gear has a striking resemblance to the modern form of Walschaerts gear minus the combination lever or crosshead component. The return crank pin was at right angles to the main crank, but as one return crank served for both fore and back gear no lead was possible.

The return crank drive was at this date not new. In 1837 Boyden, in the U.S.A., adopted a valve motion operated by a return crank, through an open link, the eccentric rod from return crank being attached to top and bottom of link as required for fore or back-gear, this link arrangement being used by Carmichael, of Dundee, on marine engines 20 years previously.

Egide Walschaerts started work as a mechanic on the Belgian State Railway at Malines; he was made shop foreman at Brussels in 1844 at the age of 24, and in October of the same year Mr. Fischer, engineer to the railway, filed for Walschaerts a patent relating to a new system of steam distribution to steam engines. The Belgian patent was dated November 30th, 1844. On the month previous Walschaerts took out a patent for the same invention in France. In 1845 Walschaerts undertook to obtain a patent in Prussia, but there is no evidence of this contract being carried out. The Belgian patent, Fig. 48, shows the link oscillating on a fixed shaft, the link having an enlarged opening in the centre and only at the extreme end of the link was the bolt, acting as a die block, without " play " or lost motion. The single eccentric rod terminated in a short tee carrying two pins; the reversing shaft, through the medium of a connecting link, raised or lowered the eccentric rod as required. The angle of oscillation of the link varied with the position of the pin in the link, and this movement was transmitted by a short arm to the combining lever operated from the crosshead of the piston rod. Fig. 49 gives a line diagram of the gear as originally patented.

23—" Engineer," December, 1912.

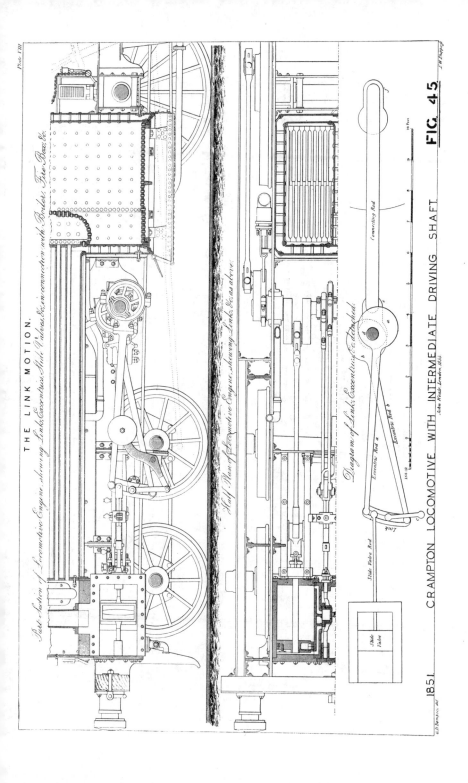

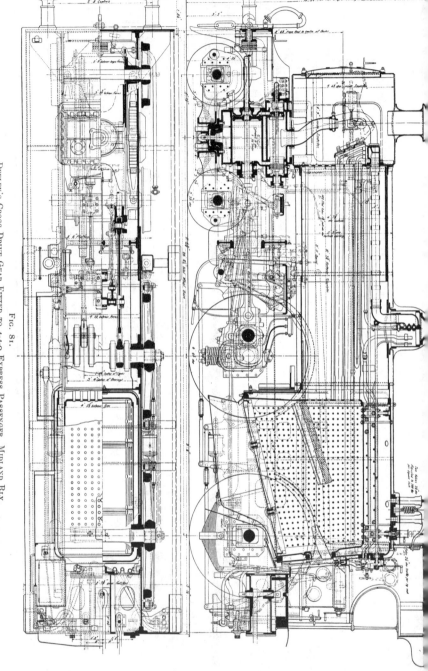

Fig. 81.
Deeley's Cross Drive Gear Fitted to 4-4-0 Express Passenger, Midland Rly.

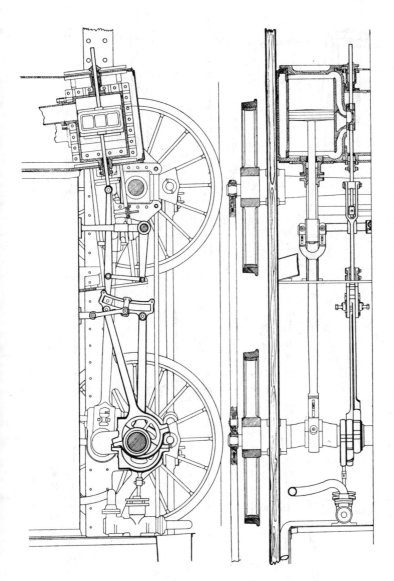

FIG. 46. GOOCH VALVE GEAR. 1843.

32 THE EVOLUTION OF LOCOMOTIVE VALVE GEARS

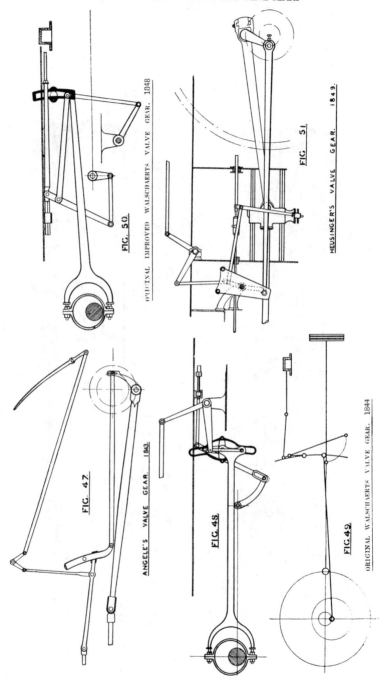

Fig. 50 shows the first improvement on this original gear as applied to Locomotive No. 98, Belgian State Railway, in 1848. The link, which was originally open at the centre, was replaced by one of equal section along its length and received its movement from a single eccentric, and except for the different placing of the link and combination lever the gear is almost as used in modern practice.

The valve gear invented by Edmund Heusinger, of Prussia (later named Heusinger von Waldegg), in 1849, was somewhat similar to Walschaerts' design as may be seen from Fig. 51,[24] a return crank was used instead of an eccentric, the combination lever was in two parts, one on either side of the crosshead and these rocked on a pivot carried by a bearing below the crosshead. The link was of a three-piece pattern carried by trunnions bolted to a cylinder bracket. In Germany it was claimed that von Waldegg was the first to invent this form of valve gear, but in 1875[25] Waldegg admitted the prior claim of Walschaerts as the first to introduce this form of motion.

Fig. 52 shows the Walschaerts gear as applied to an outside cylinder tank engine with slide valves (outside admission), and Fig. 53 gives the gear as applied to an inside cylinder passenger locomotive having piston valves (inside admission), the link, in the latter case being driven from an eccentric on the crankshaft, both of these examples being due to the late Mr. Peter Drummond on the G. & S.W. Railway in 1913.

Fig. 54 is a modern and popular design of the Walschaerts gear, the radius or valve rod being extended beyond the link and the reversing rod lifting arm, provided with a die-block, is attached to this extension as shown.

The first locomotive in this country to be fitted with Walschaerts' gear was built by the Avonside Engine Co., Bristol, in 1878, and this or a similar engine was bought by the Swindon, Marlborough and Andover Railway in 1881.[26] Beyer, Peacock and Co., built two compound engines for the Belfast and Northern Counties Railway in 1890 fitted with this gear. In England, by 1913, about 135 main line engines were fitted with Walschaerts' gear.[27]

It is generally assumed that the Walschaerts gear was introduced into the U.S.A. by William Mason in 1875, this design being shown by Fig. 55. It will be seen that the point of suspension of the radius rod was hung from a die block attached to a stationary link or guide, in which the point of suspension conformed to the curve of the link, this being claimed to produce uniform steam distribution at both ends of the cylinder.

Allan's Straight Link Motion.

Alex. Allan, Superintendent of the Scottish Central Railway at Perth, invented his " Straight Link Motion " in 1855, the gear being

---
24—" Locomotive Magazine," Vol. XXXVI, page 60.
25—" Locomotive Magazine," Vol. XXXVIII, page 313.
26—British Steam Locomotive, 1825-1925, page 224.
27—The Development of British Loco. Design, 1914, page 73. E. L. Ahrons.

34 THE EVOLUTION OF LOCOMOTIVE VALVE GEARS

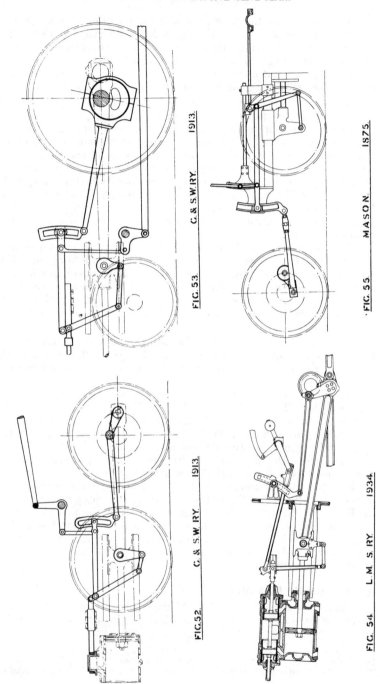

FIG. 53. G. & S.W. RY. 1913.

FIG. 55. MASON. 1875.

FIG. 52. G. & S.W. RY. 1913.

FIG. 54. L.M.S. RY. 1934.

first tried on an 0.4.2 engine of that railway, and later the Allan motion became common in this country and abroad. At present the Loch and Castle class engines on the Highland Section, L.M.S. Rly. have the Allan type of link motion. Highland Railway No. 35

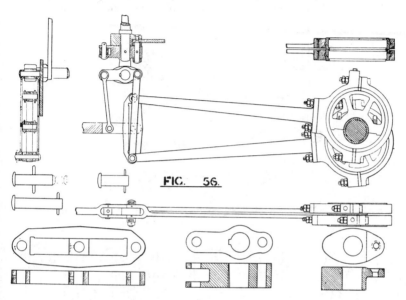

ALLAN'S STRAIGHT LINK MOTION. BEYER, PEACOCK. 1868.

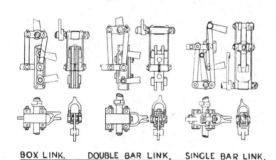

BOX LINK.   DOUBLE BAR LINK.   SINGLE BAR LINK.

FIG. 58.

(L.M.S. Rly. No. 14686) "Urquhart Castle" built in 1911, and a few Loch class engines built in 1917 were probably the last main line locomotives in this country to be fitted with the straight link motion.

Fig. 56 shows a design of Allan motion by Beyer, Peacock & Co., in 1868. It will be observed that the gear has both the features of the Stephenson and Gooch gears, these being combined, both the link and die block being moved to get the die block in required position of the link. In the straight link the curves or arcs described by the movement of the eccentrics and valve connecting rods neutralised each other and thus effect the object of the usual curve

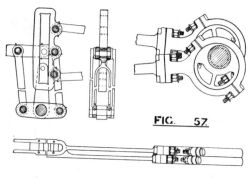

FIG. 57

ALLAN'S STRAIGHT LINK MOTION.   L.&S.W.RY.   1870.

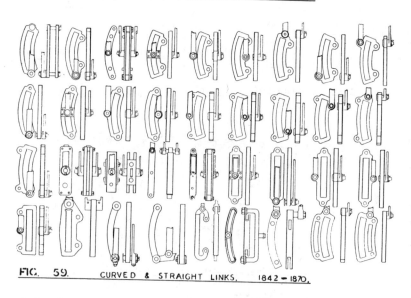

FIG. 59.   CURVED & STRAIGHT LINKS,   1842 – 1870.

of the link and die block in Stephenson or Gooch gear. In the Allan motion the movement of the link and die block was accomplished simultaneously by joining the rod corresponding to that by which the link is suspended in the stationary link motion, to an arm from the reversing shaft and connecting the valve rod to another arm from the opposite side of the reversing shaft, by which movement of the reversing shaft will give to the link and die-block movements in opposite directions. The advantages of this gear were that the motion was simple in construction, required less room than the shifting link motion, and generally the back balance or spring on reversing shaft was not employed.

In Fig. 57 is shown another design of Allan motion as constructed by C. H. Beattie on the L. & S.W. Railway in 1870. Here the reversing shaft was placed beneath the links, and the link suspension was taken from the centre of the straight link. Fig. 58 shows three different designs of straight link motion using a box link, double bar link and single bar link.

Various types of curved, stationary and straight links as used between 1842-1870 are shown in Fig. 59.[28] The chief feature in the 36 types shown is how the point of suspension varies in almost every case.

Stewart's Valve Gear.

Mr. Stewart, of Sharp, Stewart and Co., Manchester, in conjunction with a Mr. Hope, revived the principle of the eccentric band (as proposed by Williams in 1842) when he patented, in 1857, his improvements in valve gear for stationary and locomotive engines. Stewart adopted a straight link hung in a manner somewhat similar to the Gooch stationary link and connected a die block by a rod from the reversing shaft. Fig. 60 illustrates this gear as applied to inside cylinder locomotives, in which the radius rods sustained the fulcrum of the link, and the centre on which the sustaining rod oscillated, is the rod connecting this point with the motion plate of the engine. There was an arm extending at right angles from the sustaining rod of one eccentric, forming a lever or rod connecting the end of the arm with the back strap of the other eccentric. This was necessary in applying Stewart's gear to locomotive engines to keep the fulcrum of the link in the same relative position with the crank axle.

To overcome the irregularities caused by the varying position of the crank axle relative to the framing of the locomotive the arrangement made consisted of the strap of one eccentric being made to impart motion to a lever, so as to agree in time, direction and extent with the movements of a part coming from the strap of the other eccentric, these two parts being connected by a rod joined to each, so that the forces acting upon one eccentric strap were counteracted by the forces acting upon the other, the centre of the lever being practically motionless under the action of the two eccentrics and were only required to be connected to motion plate or engine framing by a rod placed horizontally to prevent the links rotating with the eccentrics and to allow for the vertical movement of the locomotive on its springs.

Link and Cam Motion.

A combined link and cam motion was patented by Uhry and Luttgens in 1866, and this American gear is shown in Fig. 61. The

---
28—Link Motion and Expansion Gears. N. P. Burgh. 1870.

ordinary link motion was connected with a pin attached to the lower arm of the rocking shaft, a second rocking shaft AA^1 was pivoted to the top pin of the main rocking shaft. The lower arm A^1 of the second rocking shaft was bent into a half circle to clear the main rocking shaft B. The second or supplementary rocker was worked by a cam and frame, the cam rod being connected to the pin C. The action of the cam was to accelerate the valve at the points of

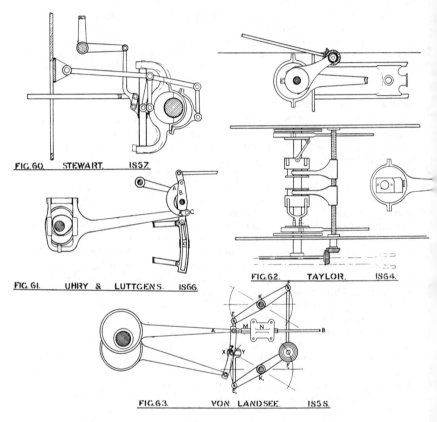

FIG. 60. STEWART. 1857.
FIG. 61. UHRY & LUTTGENS. 1866.
FIG. 62. TAYLOR. 1864.
FIG. 63. VON LANDSEE. 1858.

admission and exhaust. Its adjustment gave about 50 per cent. greater opening of steam port, the point of release was retarded 5 to 6 inches beyond the link motion, the point of compression remaining the same. Rodgers built one engine for the Central Railroad, New Jersey, with this gear.[29]

Separate Eccentric Shaft Gear.

Timothy Hackworth's "Royal George," built in 1827, had its slide valves worked from a valve shaft or eccentric shaft rotated

---

29—Development of the Locomotive Engine. A. Sinclair. Page 470.

through a system of levers from the small end of the connecting rod, although later this gear was altered and the eccentrics placed on the driving shaft. In 1864 we have this idea revived by Taylor and Dow, of Stratford, Essex, in what was called " Taylor's Shifting Eccentric Gear." It will be seen from Fig. 62 the eccentrics were mounted on a block, the block being keyed to the crank shaft, or as in figure, to an additional eccentric shaft driven from coupling rod crank pin or other convenient means. The eccentrics were so fitted as to be carried round with the block and also were controlled to slide in the direction of their diameter, being supported on either side by check plates on the block. From the eccentric straps the eccentric rods lead direct to the valve rod, as with an ordinary fixed eccentric. A little distance from the block was placed a clutch, sliding on the eccentric shaft and at the same time being carried round with the shaft. The clutch had a groove round it, to which ends of a forked arm clasped for the purpose of sliding it toward or from the eccentric. On the side next the eccentric the sliding piece had a projection which was extended towards the periphery of the eccentric near which was attached a joint pin to lugs projecting on the side of the eccentric. The traverse of the sliding piece on the shaft caused the eccentric to traverse in the direction of its diameter and to transfer the point of greatest eccentricity from one point of its circumference to another and so produce a corresponding change in the position of the valve. Instead of moving in a straight line the eccentrics could be made to move in a curved course on its block, which would be desirable, to give lead to the valve and also lap to work the steam expansively. To impart movement to the screwed shaft a hand wheel on the footplate was operated by the driver through the medium of bevel wheels as shown.

Parallel Suspension Bar Gear.

Amended and modified forms of link motions were constantly being brought out in the fifties of last century. As an example, von Landsee in Germany modified the link motion as shown in Fig. 63. When the link motion was suspended by a suspension link, the forward and backward motion of the point of suspension was a curve or arc the radius of which was the length of the suspension link. To obviate this motion and make the point of suspension move in a straight line, von Landsee adopted the arrangement as shown. The parallelogram $E.E_1$; $F.F_1$ was movable round the points $K.K_1$, the vertical links $E.E_1$ and $F.F_1$ moved on the turning of the horizontal arms $E.F.$ and E_1F_1 in a direction parallel to each other and the guide xy which was fastened to the vertical rod $E.E_1$ was provided with a slot parallel to the centre line A.B. The guide xy always occupied positions at which this parallelogram remained unaltered. In the slot of the guide xy the centre pin Z of the link moved backwards and forwards. The parallelogram which was provided at F_1 with a balance weight, was moved from the footplate through the shafts $K.K_1$. M.B. represents the valve rod and N. its guide, the motion being shown in fore gear.

RADIAL VALVE GEARS.

Hackworth's Gear.

Radial valve gears other than the Walschaerts type may be said to have begun with Melling in 1830, then followed Hawthorn in 1838, and Nasmyth in 1844. Later came Hackworth's in 1849 and an English patent of the French Engleman gear in 1859. An improved Hackworth gear was patented in 1859, a line drawing of which is given by Fig. 64. The slide valve received its motion from a single point in the mechanism of the engine, there being one eccentric for each slide valve. The eccentric had a rod whose extreme end moved in a straight path. The valve rod was attached to the eccentric and was at right angles to it; this attachment was about two-thirds along the eccentric rod from the eccentric, while the other extremity of the eccentric rod was guided by a parallel motion carried on a weigh shaft. This parallel motion acting as a fulcrum, and when the rectilinear movement ran through the centre of the crankshaft a proportional amount of the eccentrics throw was communicated to the valve, the amount being that of lap plus the lead, the extremes of the movement occurring when the engine was on dead centres, this giving the lead for either direction of motion and determined the relative position of crank and eccentric. A variable action of the valve rod was produced by causing the parallel motion or guide to move angularly on either side of the centre line as represented by a X, the port opening to steam being reduced as the line of motion approached the centre line, in which the motion of the slide valve moved only the distance of lap plus lead, the motion then being in mid-gear.

The straight slotted bar HK could be fixed in any position about centre N. The motion is in full forward gear (direction of arrow) when the slotted bar is in K_1H_1 position shown, in full back gear when in H^1K^1 and mid-gear HK. When the crankshaft S turns the centre of eccentric described the circle PFO, the block G slid up and down the slotted link, the point L described an elliptic figure and imparted this movement to the slide valve. Fig. 65 illustrates this gear as applied to a locomotive in 1868, and Fig. 66[30] shows a modern application operating Lentz valves on the Netherlands Railways in 1926.

Marshall's Valve Gear.

In 1879 F. C. Marshall, of Newcastle, patented a modified form of Hackworth's gear, as shown in Fig. 67. The crank, eccentric and eccentric rod were arranged as in Hackworth's motion, but the sliding block or guide was replaced by a swing link GH^1, this being carried by the end of the reversing rod, which put the mechanism in full gear or mid-gear according as it occupied the positions NH^1, NH_1 or NH respectively. The point G was constrained to move in the arc of a circle instead of in a straight path, as in Hackworth's gear.

30—" Locomotive Magazine," Vol. XLIV, page 218. Also November, 1924.

THE EVOLUTION OF LOCOMOTIVE VALVE GEARS

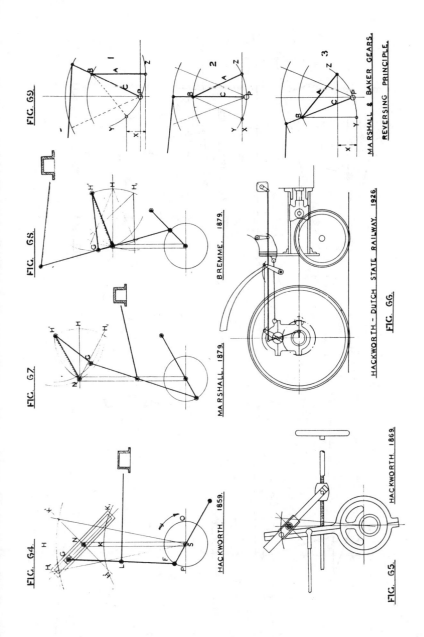

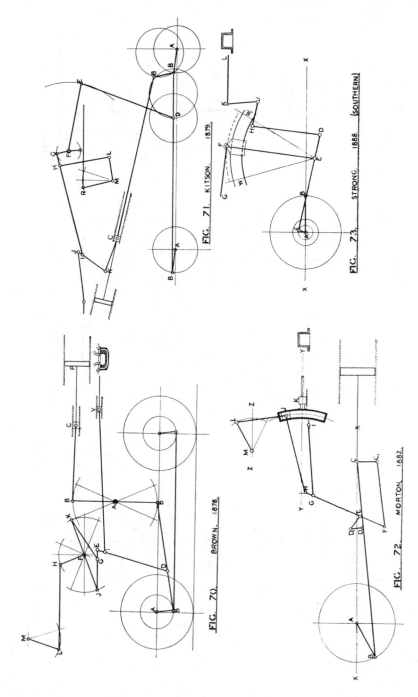

Bremme's Valve Gear.

Also in 1879 C. A. Bremme, of Liverpool, patented another form of Hackworth's gear, Fig. 68, this being chiefly used for marine purposes. The point of attachment of the valve rod in this gear was beyond that of the swinging link GH instead of between the latter and the eccentric, as in two previous gears, and in this case the centre line of the eccentric and crank coincide.

Principle of Reversing Radial Gears.

Fig. 69[31] shows the principle of reversing gears of the Hackworth pattern and of the modern Baker and Southern gears, in which the reversing member C and the radius bar A provide the means for reversing the direction of motion of the engine. In the top figure (1) the reversing lever or member C is in full forward position; the lower end of radius bar A moves in the arc of a circle YZ, whose centre is at B, the point of attachment of the upper end of the radius bar A. The lower end of the radius bar, and the end of eccentric rod attached to lower end of radius bar will have a vertical movement equal to the distance X. As eccentric rod is drawn to the left its end is raised along the arc YZ and as it is moved to the right this end is lowered along the same arc. As the reversing lever C is drawn towards its central or vertical position as in the middle figure (2) the arc YZ will be brought nearer to the horizontal position, and a considerable reduction will take place in the amount of vertical motion X imparted to the end of eccentric rod, the extreme position Y and Z of the lower end of the radius bar A are at the same level so that the vertical movement of X is zero. As the reversing lever is pulled back to full back gear position as in lower figure (3) the arc YZ is inclined in the opposite direction to view (1). As the eccentric rod is drawn to the left its end is lowered along the arc YZ and when moved to the right the end of eccentric rod is raised, producing the opposite effect to that produced when the reversing lever was in full forward position. This reversal of movement of the eccentric rod produces the means of reversing the engine, just as lowering or raising the link in Stephenson's link motion or lowering or raising the die block in Walschaerts gear. In order to overcome the variation in the valve travel due to angularity of the various parts in the Baker gear (Fig. 88) a bell crank is placed between the eccentric rod and the valve rod to give an equal steam distribution for both forward and backward strokes of the piston.

Brown's Valve Gear.

Mr. C. Brown, of the Swiss Locomotive Works, introduced different forms of radial valve gears from 1877, varying from the Hackworth to the Joy types, although his latter pattern preceded that of David Joy.[32] Fig. 70 gives an arrangement of Brown's

---
31—Locomotive Valves and Valve Gears. Yoder and Wharen. 1917.
32—" Engineering," Vol. XXX, page 271.

valve gear for small locomotives. In this case the motion of the piston was not communicated direct from the crosshead through the medium of the connecting rod to the crank pin, but through an upper and lower connecting rod with a rocker between. The valve was driven from a point D on the lower connecting rod through a lever DE whose upper extremity E is fulcrumed on the describing point of a slider crank parallel motion KJGE. The describing point E moved in a path more or less curved, and in addition, the axial link JK was arranged to be held in any position about a fixed centre F giving a mechanism suitable for the intermediate linkage of a valve gear of the Hackworth type. If E moved for a limited range in a straight line passing through F, we have the Hackworth gear over again.[33] A constant lead of the valve, with a quick cut-off and a quick release were obtained. The motion of the valve was composed of two movements, one being designed to place the valve in position for the beginning of each stroke and the other to give the required travel of the valve. The motion taken off the lower connecting rod at D, in which the vertical movement was about $\frac{3}{8}$ of the traverse axis of the engine stroke. The other end E of the lever was joined to the valve by the radius rod IV, connected to a point near to its fulcrum. This was proportioned so that the valve was set for constant lead. The lever DE also supplied the required travel of the valve. The fulcrum E was connected to a bar at G which was held at one end by the link GK pinned to the reverse lever JK, and at the other end by a sleeve at J, also attached to the reverse lever, in which the end of the bar was free to slide to suit the varying angles of the link GK. The travel of the valve was obtained by the decomposition of the vertical movement of the lever DE which at the upper end was deflected laterally by the oblique play of the lever GK, the lateral component of the motion being communicated to the valve through the radius rod IV. The greater the obliquity of the rod GK the greater the travel of the valve.

Kitson's Valve Gear.

Another form of valve gear used on small and tramway engines was that patented by J. H. Kitson in 1879 as given by Fig. 71.[34] This was a modification of the Walschaerts type; the drive for the link being taken from the coupling rod, the rod DE taking the place of the eccentric rod, and was joined to an arm EF on the link. The point of attachment D of the rod DE on the coupling rod BB described a complete circle.

Morton's Valve Gear.

In 1882 D. Morton patented a valve gear based on the Hackworth pattern, and this also was principally used on small locomo-

33—Prof. A. Maclay, Glasgow Technical College.
34—" Engineering," Vol. XXX.

THE EVOLUTION OF LOCOMOTIVE VALVE GEARS 45

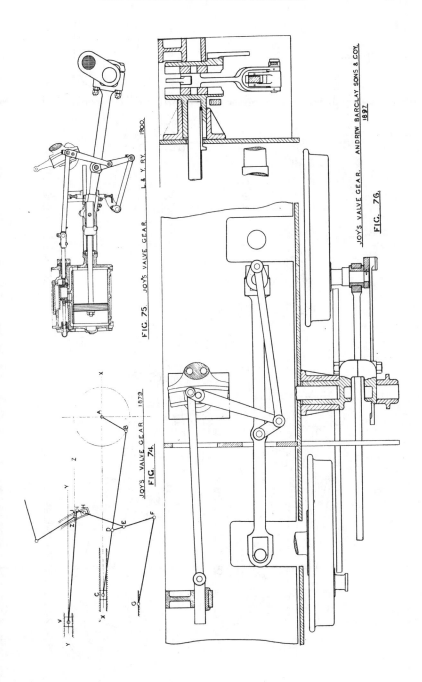

FIG. 74. JOY'S VALVE GEAR 1879.

FIG. 75. JOY'S VALVE GEAR L. & Y. RY. 1900.

FIG. 76. JOY'S VALVE GEAR. ANDREW BARCLAY SONS & COY. 1897

46 THE EVOLUTION OF LOCOMOTIVE VALVE GEARS

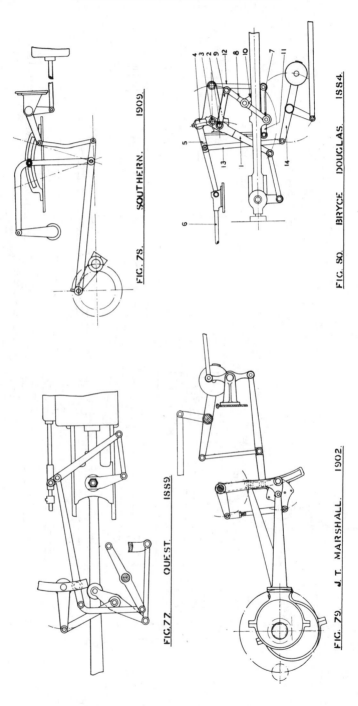

tives, although both the North British and Great Eastern Railways gave this gear a trial. In Fig. 72, CC_1 was a rigid arm attached to the engine crosshead, and DD_1 was a similar arm attached to the connecting rod; each of these in turn arms carried a swinging link, C_1F and D_1E respectively, and these carried one end of the valve lever FH. The motion of the valve lever was made determinate by being guided towards its other end by an anchor link G1, and its extremity H worked the slide valve through a radius rod HJ and the link on the end of valve spindle KY. The position of the die block J in the link was regulated from the shaft $M.^{35}$ In describing this gear it would appear to come under the Walschaerts class, as it has both a crosshead and connecting rod arrangement for working the valve gear and the valve or combination lever is attached to a radius rod which is centred on a die block in a curved link. The affinity is more apparent than real as there is an inversion in the relation between the radius rod and curved link as driver and follower, as in the Morton gear the radius rod is the driver and the link has no oscillating movement but was part of the valve spindle. In Walschaerts' gear the crosshead connection and eccentric rod connections have separate functions, but in Morton's gear the valve travel was always the resultant of the motion of both connections, also the curved link in Morton's gear had to accommodate itself to the radius rod by shifting along the line of the valves motion while the radius rod swung about the extremity of the valve lever as centre.

Strong's Valve Gear.

Strong, in the U.S.A., patented in 1888 a form of Hackworth gear as shown in Fig. 73. AB is the crank and AC the eccentric. CD, the eccentric rod, was fulcrumed on the lower end of the swing link EF. The pin F was attached to a die block which could be set at any position on a curved guide WW_1 by the reversing lever FG. The other end D of the eccentric rod CD was connected with a bell crank lever HJK having a fixed centre pin J, the valve V being driven from the rod HL. The eccentric was set so that the angle ACB equalled 90 degrees. The figure shows the gear in mid position. For full gear the pin F was moved forward or backward to W or W_1 for fore and back gear respectively. This gear was a Bremme variety of the Hackworth type, the die block F and curved link WW_1 taking the place of the anchored swing link in Bremme's gear. Other forms of Hackworth's gear were introduced in the U.S.A. during the eighties, these being generally of the Marshall type.

Joy's Valve Gear.

In the late sixties David Joy worked on his radial valve gear, and in 1879 he patented his now well-known Joy valve gear, while he was with the Barrow Shipbuilding Co. The Joy gear was first

---

35—Prof. A. Maclay, Glasgow Technical College.

used on a locomotive in 1880 on the Furness Railway and later with the L. & N.W. Rly., this lead being followed by other railways at home and abroad, and by 1897 there were 3,345 locomotives in England fitted with Joy's gear,[36] and the last main line locomotives to be fitted with this gear were Beames' 0.8.4 tank engines built for the L.M.S. Rly. in 1923.[37]

The elements of the gear are given in Fig. 74 where XX is the centre line of the cylinder, CB the connecting rod, AB the crank and VI the valve rod. Swinging about the centre G is the anchor link GF connected by a stirrup link FD to a pin D on the connecting rod, while in the link DF is a pin E joined by the swing link E1 to the pin 1 on the valve rod V1, the last mentioned link moving on H as a fulcrum. The crosshead C moved in a straight line XX while the crank pin B moved in a circle with AB as radius. The intermediate point D in the connecting rod moved with an oval motion. The point F moved in an arc of a circle with the radius GF and the point E moved in a path more or less oval in shape, but flattened on the top and bulging at the bottom. The axis of this curve is horizontal but at the point H in the swing link E1 the curve of motion was converted so that the axis was more or less vertical. The point H was the centre of a die block that slid freely in a radial guide, the vertical motion was thereby eliminated as regards the rod VI, the horizontal portion of the motion working the valve rod and moving the valve.

The mechanical relation between this gear and Brown's is the same as that between Bremme's and Marshall's. The valve lever of the two former takes the place of the eccentric rod in the two latter; in each the first point drives and the second point guides, and the third moves with the resultant motion of the first and second.

Fig. 75 illustrates Joy's gear as used about 1900 and Fig. 76 gives another arrangement for outside cylinders, in which the anchor link was attached to a small return crank in line with the main crank, this being used on a small locomotive by A. Barclay and Sons, Kilmarnock, in 1897. A similar arrangement was previously used on 2.8.0 type locomotives on the Buenos Aires and Pacific Railway in 1884.

Bryce Douglas Valve Gear.

In 1884 Bryce Douglas, of Ardrossan, patented a form of valve gear much more complicated than Joy's. In addition to being used on marine engines the gear was tried on the G.E. Rly. and also on the Caledonian Railway Engine No. 124 (Eglinton) exhibited at the Edinburgh Exhibition during 1886. As this gear became infamous with its repeated failures its description from the original patent may be of interest.[38] In Fig. 80 the beam or combination lever (1) was mounted on a stationary fulcrum (2) and

36—" The Development of British Loco. Design." Ahrons. 1914.
37—" Locomotive Magazine " XLVI, page 156.
38—Patent No. 4958, of 1884

connected at one end to the engine crosshead, so that the lever oscillated in time with the strokes of the piston. On a pin (3) on this lever was mounted a curved link (4) on which was fitted a sliding block that was worked by a rod (5) to the slide or valve rod (6). A pin (7) was connected to the connecting rod by a link (8) to the middle joint of a pair of toggle links (9 and 10) of which (9) was pivoted on the combination lever (1) and (10) was joined to the radius rod (11). This radius rod was pivoted on a stationary fulcrum and a pin on it was connected by a link (12) to an arm of the curved link (4). By means of these connections the link (4) received a motion corresponding nearly with that of the link usually worked by two eccentrics. The rod (5) being linked by a rod (13) to an arm (14), by moving this arm the block at the lower end of (5) could be slid along the curved link (4) and thus the valve rod (6) could be made to operate the engine for fore and back gear and for various degrees of cut-off.

" Ouest " Valve Gear.

In the late eighties of last century there appeared on at least two French railways a form of valve gear in which the lap and lead movements of the valve was obtained from the crosshead components as in Walschaerts' gear, but the oscillating action of the link was obtained, as shown by Fig. 77, from a point on the connecting rod as in Joy's gear, in which the swing link driven from the stirrup link worked a horizontal arm on the back of the expansion link.

Southern Valve Gear.

The Strong (U.S.A.) type of valve gear was " invented " or re-introduced over 30 years ago by W. S. Brown on the Southern Railway (U.S.A.) and is known as the Southern locomotive valve gear, as shown by Fig. 78. This gear is also suggestive of the Marshall type, the chief difference being that instead of using the reversing member a stationary curved guide is used, the arc of which has a radius equal to the length of the vertical radius bar. The die block attached to the radius bar and also to the reversing lever, has the same effect on the gear as the reversing member has in the Marshall and Baker gears.. The transmission to the valve rod is through the medium of a bell crank. The lead is constant and is determined by the ratio of the two arms of the eccentric rod.

" J. T. Marshall " Valve Gear.

In 1902 Mr. J. T. Marshall, of Leeds, introduced a novel form of valve gear shown by Fig. 79,[39] this gear having for its object a quick travel of the slide valve at the points of opening and closing and a slow movement when steam and exhaust ports were full open. The curved reversing link was of the Gooch type, reversal being obtained by shifting the valve rod. One eccentric rod was at 180

39—" The Development of British Loco. Design." Ahrons. 1914.

degrees to the crank and was attached to the centre of the reversing link. This eccentric had a travel of twice the outside lap and lead and its action was to move the expansion link bodily to and from the crankshaft (as the combination lever did in the Bryce Douglas gear). The end of this eccentric rod, in addition to being attached to the link, was also attached to a swinging rocker arm, this arm, suspended from a cross shaft, carried the trunnion on which the reversing link worked. The oscillation of the link was obtained from another and larger eccentric, fixed at about 90 degrees behind the crank and its eccentric rod actuated the vertical arm of a bell crank lever rocking on the same cross shaft as the suspension links. The horizontal arm of the bell crank lever extended backward and was connected by means of an adjustable coupling link to an arm formed on the trunnion bracket of the reversing link. In this gear we see a marked similarity to the much-abused Bryce Douglas gear, only Marshall used eccentrics. The slow movement of the valve during the exhaust period was produced by the combined action of the two eccentrics on the floating expansion link, the movement of the one neutralizing that of the other. This gear was used on the G.N. Railway and on a larger scale on the G.S. & W. Railway, Ireland, and in 1930 was fitted to the inside cylinder of the three cylinder 0.8.0 locomotives S. Rly.[40] J. T. Marshall's gear should not be confused with the earlier Marshall gear, by F. C. Marshall, of Newcastle, in 1879.

Cross Drive Valve Gear.

In 1830 we have seen that Melling improved his original gear by operating the slide valve from the connecting rod of the contrary cylinder. Richard Roberts, of Manchester, in 1832, invented a valve motion somewhat similar in principle. Later, James Nasmyth, in 1844, worked the link on his expansion valve gear from the crosshead of the opposite cylinder. In the late sixties M. Stevart,[41] in Belgium, used an amended form of Walschaerts valve gear in which the combination lever was of the usual type but the expansion link was driven by connections through cross shafting with a rocking lever from the opposite cylinder, i.e., the movement usually taken from the return crank or eccentric to operate the link, was instead, taken from the opposite crosshead.

Mr. R. M. Deeley, of the Midland Railway, designed, in 1887, a valve gear similar in principle to that of Stevart. The oscillation of the expansion link for the right cylinder was derived from the crosshead of the left piston and vice versa. The left expansion link was placed in front of the right hand link and from the position of the two arms of reversing lever the die block of one link was at the top when the other was at the bottom, Fig. 81. It was not till 1907 that practical use was made of Deeley's gear when ten 4.4.0 locomotives were fitted between 1907 and 1909.

40—" Locomotive Magazine," Vol. XXXVI, page 404 and Supplement.
41—" Locomotive Magazine," Vol. XLII page 353.

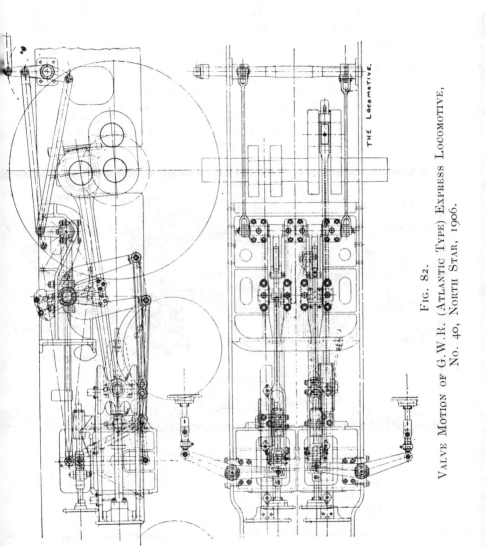

Fig. 82.

Valve Motion of G.W.R. (Atlantic Type) Express Locomotive, No. 40, North Star, 1906.

Mr. Churchward, in 1906, introduced on the G.W. Rly. a four cylinder Atlantic type locomotive, " North Star "[42] with a gear similar to Deeley's. In the Stevart and Deeley gears the centres of one link were some distance behind the other, this requiring radius or valve rods of different lengths, but in the Churchward gear, Fig. 82, both radius rods were of equal length. The outside cylinder piston valves were operated by a horizontal rocking shaft driving the outside valves from the front of the cylinders. In this gear both radius rods were of equal length, mutual freedom being secured by curving the link operating arms and also making one of the driving rods from the crosshead longer than the other. To avoid one die block being at the bottom of link when the other was at the top the reversing arm had an upper and lower arm, from

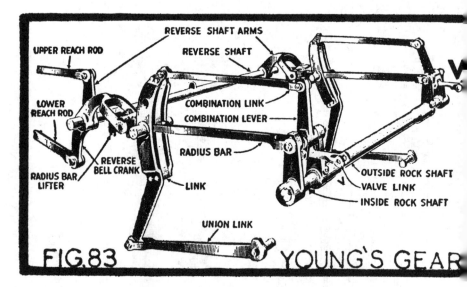

which extended rods to two secondary shafts, one on either side of the engine, from which projected arms carrying blocks sliding in slotted extensions of the radius rods. This engine was later converted to the 4.6.0 type, but the valve gear remained till 1930, when the engine was again rebuilt.

Valve gears of the above type are known in the U.S.A. as the Young locomotive valve gear, O. W. Young introducing this gear in America along with a special type of piston valve. From Fig. 83 it will be seen that a union link transmits the movement of the engine crosshead to the expansion link fixed on central trunnions. A bell crank to the rear of the expansion link carries a movable die block connected to the radius bar, this radius bar being connected to a cross rocking shaft to give the valve move-

---

42—See " The Locomotive Magazine," Vol. XXXVI, page 274.

ment to the opposite valve. An upper and lower reach rod connects to the reversing bell crank, this bell crank shaft extending across the engine frame. From the top of the expansion link a combination link connects to an inverted combination lever, at the lower extremity of which is the connection V to the valve spindle. Each crosshead produces a valve movement equal to the lap and lead on its own side and the valve travel for the opposite side. The rock shaft extending across the engine consists of an inner part which acts for one side of the engine and the outer part or tube acting for the other.

Amended Walschaerts Gears.

Walschaerts valve gear has been used with various amendments and modifications in attempts to improve the steam distribution in the cylinders.

Stephens.

In 1885 A. J. Stephens, in the U.S.A., used the combined movement of a single eccentric and combination lever from crosshead to impart movement to a wrist pin attached to the valve rod. By this arrangement the valve opened quickly for admission and remained full open for an instant before rapidly closing. The lap and lead movement and variable valve travel being combined.

Baguley.

The Baguley valve gear used on small locomotives in 1894 was similar to Stephen's gear, but the wrist pin was in the form of an eccentric trunnion.

Jones.

This was an American gear of the Walschaerts type, having the features of variable lead. An additional link having a curved slot was connected to the reversing gear; in this slot worked an extension of the combination lever. The die block at top of combination lever was restrained from horizontal movement by a link pivoted to the engine frame. When the reversing gear was moved, the die block in the expansion link and the additional link were both moved at the same time, so that their combined movements gave a variable lead which increased toward mid-gear as with Stephenson's link motion.

Helmholtz.

This gear consisted of making the link straight, and the radius rod was connected to the lifting link instead of to the link block. The curving of the link was compensated for by the reversing shaft or lifting arm fulcrum being located in a given position above the link, so that the locus of the suspension point of the lifting link forms an arc of a circle with its chord perpendicular to the centre line of the radius rod in its centre position. The radius of this arc bears the same relation to the length of the radius rod as the distance of the radius rod connection above the link block bears to the length of the lifting link, result being that this connection is moving in an arc with a radius of the length of the radius rod and the same motion obtained as is in the Walschaerts gear.

Beames.

In the Beames modification of the Walschaerts gear the combination lever was driven from a forward extension of the coupling rod and the expansion link from a return crank on the same coupling pin. Fig. 84 shows this gear as fitted to the inside cylinder 4.6.0

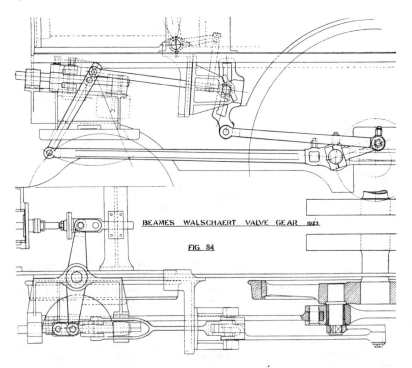

BEAMES WALSCHAERT VALVE GEAR 1923.

FIG. 84

locomotive, "Prince of Wales," on the L.M.S. Rly., in 1923. The combined motion of the combination lever and link being transmitted through a horizontal rocking arm to the piston valves between the engine frames. Golsdorf, in Austria, also used this form of Walschaerts gear.[43]

Stephenson=Molyneux System.

A recent modification of Walschaerts gear used on the Buenos Aires and Pacific Railway and made by Messrs. R. Stephenson and Hawthorns, Ltd., is that known as the Stephenson-Molyneux

---

43—"Engineer," Vol. LXXXVIII, page 318.

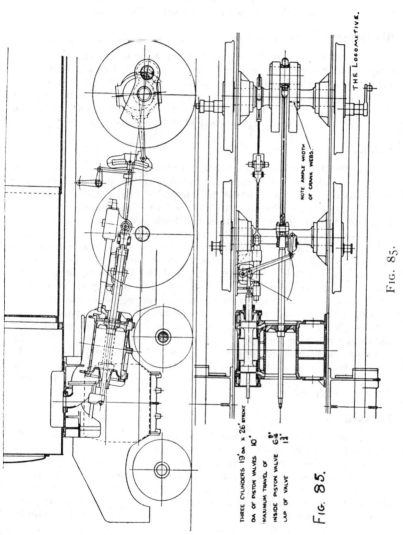

FIG. 85.

WALSCHAERTS VALVE GEARS: STEPHENSON-MOLYNEUX SYSTEM.

System,[44] and is illustrated in Fig. 85. This system permits the valve steam chest to be located above, below or at the side of the cylinder, without using rocking shafts, levers or unsupported offsets in the rods or jaw ends, allowing certain members of the gear to operate in any desired plane, inclined or horizontal.

Bonnefond.

Amended forms of Walschaerts gear have been used to operate separate admission and exhaust valves. In the French Bonnefond gear [45] each cylinder had two admission and two exhaust valves, the reversing link motion being taken from a cam on the end of the main crank pin. The point of cut-off was independent of the reversing link, by a spiral cam adjusted by means of a bevel pinion.

Berthe.

M. Berthe, of the French State Railway, in 1900, introduced a gear in which the main admission piston valves worked above the main piston and the exhaust piston valve below the piston, the exhaust valve travel being constant for any position of cut-off. In Fig. 86[46] the motion is shown in full back gear. The main link N was driven direct from the fly crank and the radius rod P from this link was connected to a rocking lever M. A short link S also connected the radius rod to the exhaust valve spindle T. The link N was only concerned with reversing, and the die block in this link had only the three positions for fore, mid and back gear, controlled by the upper of the two reversing screws on engine footplate. The rocking lever M had its upper arm connected to the expansion link G through the driving rod K. This in turn was connected through the radius rod J to the upper end of the combination lever R. The combination lever was connected to the admission valve spindle L. Cut-off was controlled by the lower reach rod C attached to the lifting rod F and the extension of the expansion radius rod J behind the expansion link G.

Marshall.

A gear of somewhat similar design was patented by J. T. Marshall, in 1926, and fitted to a 2.6.0 locomotive on the Southern Railway.[47] Fig. 87 shows that the main features of this gear were of the Walschaerts type. The radius rod N was connected to the combination lever at O below the connection P with the main valve spindle T. The pin P projected outward to connect a sliding bar U which was supported by guides, and at the front of the cylinder was joined by the curved rod W to the rocking lever X. The lower end of this lever was slotted and by means of a die block imparted

---

44—" Locomotive Magazine," Vol. XLV, page 155.
45—" Engineering," Vol. XLVIII, page 710.
46—" Locomotive Magazine," Vol. XLV, page 24.
47—" Engineer," 16/11/1928. Also " Locomotive Magazine," Vol. XLV, page 25.

THE EVOLUTION OF LOCOMOTIVE VALVE GEARS 57

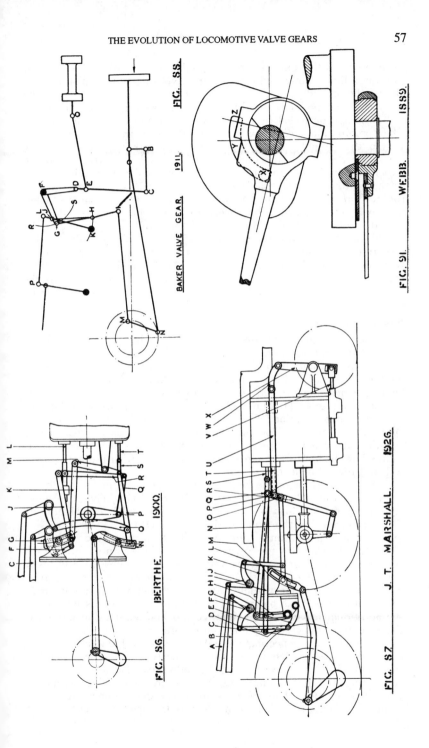

its motion to the exhaust valve spindle V beneath the cylinders. The main link L imparted a constant travel to the main piston valve above the cylinders and also for the exhaust valves below the cylinders, the position for fore, mid and back gear of the radius rod N being controlled by the lifting rod M and upper reach rod A. The combination lever R was extended above the valve spindle O (as in the Guinotte gear, Fig. 32) to the point Q where it was pinned to the rod K, the other end of rod K being attached to the upper end of the expansion link H, the expansion link being fixed at the bottom end and was therefore oscillated by the movement of rod K from the top of the combination lever. On the same bracket that held the expansion link were fixed the bearings for the two armed transfer shaft F. By means of the lifting rods D and E, the die block of the expansion radius rod I was held in any position required for cut-off, the weigh bar shaft G and the lower reach rod B being provided for this purpose. The expansion radius rod J was attached to an expansion valve spindle S which entered the steam chest above the main piston valve, this valve being of special design.

Baker Valve Gear.

The valve gear patented in the U.S.A. by Mr. A. D. Baker in 1903 and initially used on traction engines was originally known as the Baker-Pilloid gear, its first application to a locomotive being on the Toledo, St. Louis and Western Railway in 1908, the gear being further improved in 1911. The Baker gear is really a form of Marshall's gear (J. Marshall, Newcastle, 1879) with the lap and lead movement obtained from a crosshead combination lever as in Walschaerts gear. The link and block of the latter gear being substituted for a valve gear frame which contains the reverse yoke, radius bar, bell crank and gear connecting rod, with the usual reach rod and reversing levers.

A line diagram of the Baker gear suitable for inside admission s shown in Fig. 88. The radius or valve rod EO received the lower end of the vertical arm of the bell crank lever GFD combined with that produced by the combination lever DEC. The reversing mechanism is placed on a frame with the bell crank, the radius bar JH being attached to the reverse yoke. The eccentric crank MN is placed at 90 degrees behind the main crank, consequently, as in early types of valve gear, the motion from the eccentric crank alone would require a valve without lap or lead and would admit steam for a full stroke of the piston. The motion from the crosshead gives the additional travel for the required lap and lead. The motion of the eccentric crank is transmitted through the eccentric rod to the foot of the gear connecting rod GHI, the point J on the reversing bar is fixed for any definite position of the reversing lever, therefore the point H at the lower end of the radius bar JH attached to and supporting the gear connecting rod is free to swing about the point J as centre. The upper point of the gear connecting rod G, being attached to the bell crank horizontal arm, is therefore constrained in its motion and will swing in the arc rs with the pivot F as centre.

As the foot of the gear connecting rod is moved backward and forward the gear connecting rod GH1 and radius bar JH, swinging about their centres H and J will cause the end G of the bell crank lever to be moved up and down in the path rs. The lower end of the bell crank being attached to the combination lever at D and swinging about the pivot F as centre, will cause the upper end of the combination lever to swing back and forth and transmit the required motion to the valve through the valve rod EO. Reversing is accomplished by bringing the radius bar bearings in the reverse yoke past the central position away from the bell crank, thus changing the arc of the radius bars. (See Fig. 69.) The position of the reversing yoke governs the amount of travel and also the percentage of cut-off, as the reverse yoke is brought nearer to its mid position the arc of the radius bar JH becomes more horizontal and produces a less upward and downward movement of the horizontal arm FG of the bell crank, thus decreasing the valve travel.

Fig. 89 gives a recent design of the Baker gear made in this country for a Dominion railway, and Fig. 89a shows the valve ellipse for this gear. Fig. 90 illustrates an experimental form of gear by Andrew Barclay and Sons, Kilmarnock.

Locomotives with more than Two Cylinders.

A locomotive fitted with three high pressure cylinders was built by R. Stephenson and Co. in 1846. The outside cranks were on the same centre and were 90 degrees in advance of the inside crank. The left hand eccentrics operated the valves of the inside cylinder and the right hand eccentrics the valves of both right and left outside cylinders through the medium of a cross shaft.

Webb.

In 1889, F. W. Webb, on the L. & N.W. Railway, re-introduced the loose eccentric to work the valve of the L.P. inside cylinder of his three cylinder compound locomotives, there being a separate reversing gear actuating the H.P. cylinders. The loose eccentric took the place of the Joy type motion previously employed for the L.P. cylinder.[48] In Fig. 91 the pin X fitted in the crank web worked in the circular slot Y in a plate attached to the eccentric. On reversing the H.P. valve gear the engine moved in the opposite direction and the pin X moved through the slot to the other end Z and pushed the eccentric sheave round in the opposite direction, thereby reversing the action of the steam in the L.P. cylinder. The point of cut-off in the L.P. cylinder was constant between 70 and 75 per cent.

Smith and Deeley.

In the Smith system of three cylinder compound locomotives used on the Midland Railway in 1901, the H.P. and L.P. link

---

48—The British Steam Locomotive. Ahrons. Page 247.

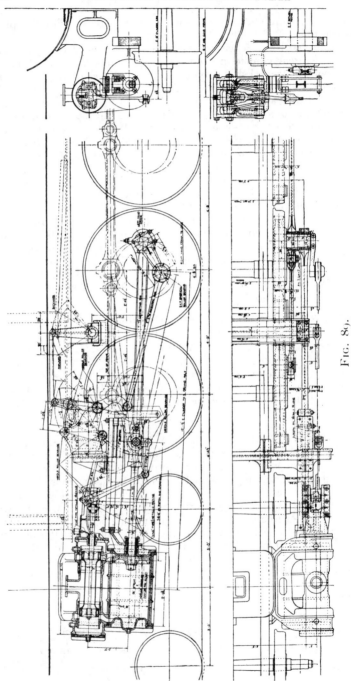

Fig. 89.
Baker Valve Gear.

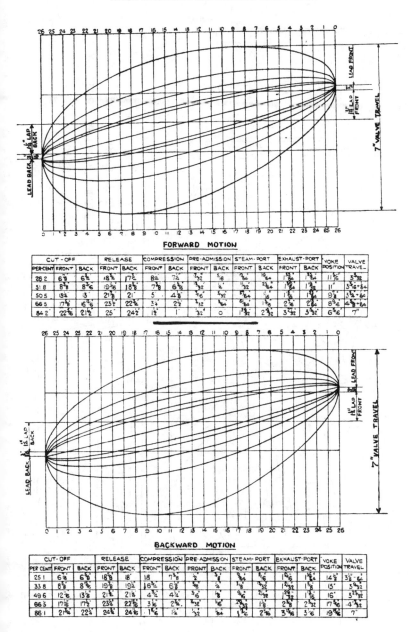

Fig. 89a.
Valve Events of Baker Valve Gear.

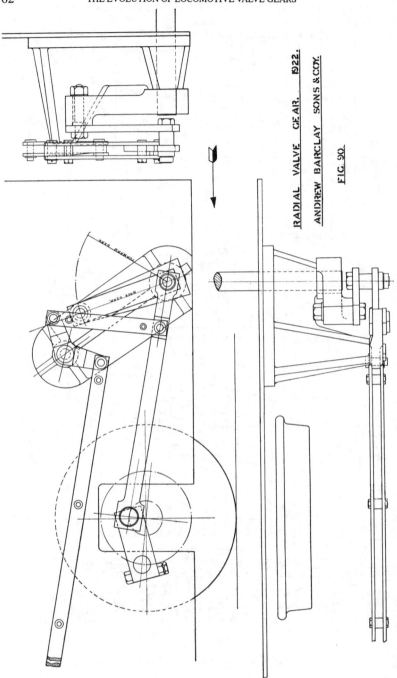

RADIAL VALVE GEAR. 1922.
ANDREW BARCLAY SONS & COY.
FIG. 90.

motion was independent, but three years later Mr. Deeley modified this type of engine with one reversing lever actuating the three sets of link motion, all three cylinders having an equal cut-off, the special type of Deeley[49] regulator valve controlling the action of the steam in the cylinders. The outside cylinder crank pins on these engines are set at 90 degrees, the L.P. inside cylinder crank pin bisecting the obtuse angle between them.

Manson.

In 1897 Mr. Jas. Manson, Locomotive Superintendent of the G. & S.W. Railway, built in Kilmarnock Works the first four high pressure cylinder locomotives in this country. The inside cylinder cranks were at 90 degrees to each other and the outside cylinder crank pins were at 180 degrees to the inside cylinder crank pins, one set of link motion working the slide valves for two cylinders. The inside cylinder slide valves were positioned vertically, between the inside cylinders, and the outside slide valves horizontally above the outside cylinders, the latter valves being operated by a horizontal rocking shaft.

This engine was re-designed by Mr. R. H. Whitelegg in 1922, in which one piston valve served two cylinders, the cylinders being cross ported. The Midland Railway four cylinder 0.10.0 banking engine, built by the late Sir H. Fowler in 1920, also employed the cross porting system.

Dendy Marshall.

In 1915 the L. & N.W. Rly. converted a two cylinder 4.6.0 locomotive into a four high pressure cylinder locomotive, and to avoid the use of four sets of valve motion or rocker arms and to avoid cross porting, the Dendy Marshall system of having two piston valves on one spindle was used.[50] There were two pistons on the valve spindle each with two heads. The front head of the front piston valve served the front port of the outside cylinder and the back head of the same piston valve served the back port of the outside and front port of the inside cylinder. The front head of the back piston valve served the back port of the inside cylinder, which was placed further back than the outside cylinder. To balance the piston valve the back piston valve had a dummy head at the back.

Holden.

Holden's three-cylinder 0.10.0 tank engine, built at G.E. Rly. Works in 1902, had one set of separate valve gear for each cylinder, the cranks being 120 degrees apart. Later three cylinder engines built for the G.E. Rly. in 1907, and by the N.E. Rly. in 1909, 1910 and 1913, were also on the same principle.

49—See " Journal of Inst. of Loco. Engineers," No. 93, page 66.
50—" The Engineer," September 24th, 1915.

Maunsell.

A number of three cylinder engines on the Southern Rly., designed by Mr. R. E. L. Maunsell, have a modified form of Walschaerts gear for the inside cylinder in which a second eccentric at right angles to the usual eccentric, supplies the movement normally derived from the crosshead by a long combination lever. In this case the combination lever is short and the lower end is connected to the rod of a second eccentric.

Conjugated Valve Gears.

The method of operating the valve of the centre cylinder of a three cylinder locomotive, having cranks at 120 degrees apart, from the outside cylinder valve gears originated with David Joy in 1884, who applied a simple sway beam, or lever, to operate one valve of triple expansion verticle marine engines from the valve gears of the other two valves (Patent No. 14107 of 1884).

In 1909 Mr. H. Holcroft devised an arrangement for locomotives in which the valve of the third cylinder, inside the frames, was actuated by levers in a horizontal plane driven from the valve gears of the two outside cylinders (Patent No. 7859 of 1909). This form of gear was the first practical application on locomotives, being adopted in Germany during 1914-15 on 4.6.0 type engines, and also in a modified form on heavy freight engines which embodied the use of rocking shafts to transfer the motion to a vertical plane.

In 1915 Mr. H. N. Gresley patented two forms of valve gear for three cylinder locomotives (No. 15769 of 1915), one of which was applied to a 2.8.0 type locomotive in 1918 (see "Locomotive Magazine," 15/10/18), the various levers and rocking shafts for operating the centre valve being placed behind the cylinders.

In 1920 the simple arrangement shown in Fig. 92 was first adopted on a 2.6.0 type locomotive and in subsequent years it was extended to a number of different types. The arrangement was also incorporated in three cylinder engines built in the U.S.A., Australia, South Africa and other countries. The extension from the right valve spindle is coupled to the horizontal two-to-one rocking lever (10) pivoted at A. The end of the shorter arm of this lever carries an equal armed floating lever (9), the end of which is joined to the left hand valve spindle extension and to the centre valve spindle extension respectively. The movement of the inside valve spindle is therefore derived from the combined movement of the extensions from the right and left valve spindles, and the cranks being 120 degrees apart, a steam distribution is obtained practically correct for the inside cylinder.

Another form of three cylinder gear, devised by Mr. H. Holcroft, was applied to a 2.6.0 type locomotive on the S.E. & C. Rly. in 1922, and to a second engine, a 2.6.4 type three cylinder passenger tank, in 1925.

Also in 1922, W. Pickersgill, of the Caledonian Railway, built four 4.6.0 three cylinder passenger locomotives with a valve gear as shown in Fig. 93. A and B were the left and right combination

THE EVOLUTION OF LOCOMOTIVE VALVE GEARS 65

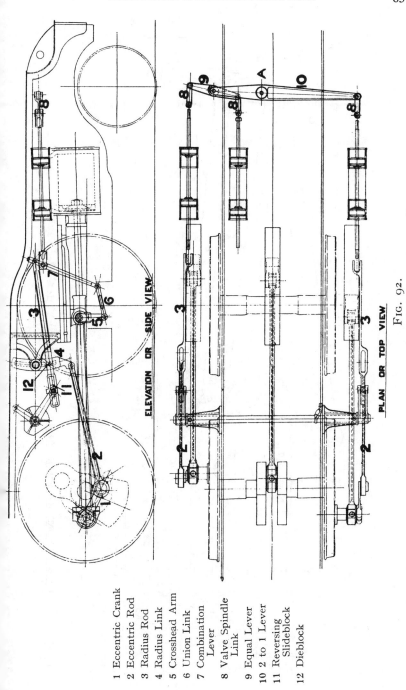

Fig. 92.
GRESLEY THREE-CYLINDER VALVE GEAR.

1 Eccentric Crank
2 Eccentric Rod
3 Radius Rod
4 Radius Link
5 Crosshead Arm
6 Union Link
7 Combination Lever
8 Valve Spindle Link
9 Equal Lever
10 2 to 1 Lever
11 Reversing Slideblock
12 Dieblock

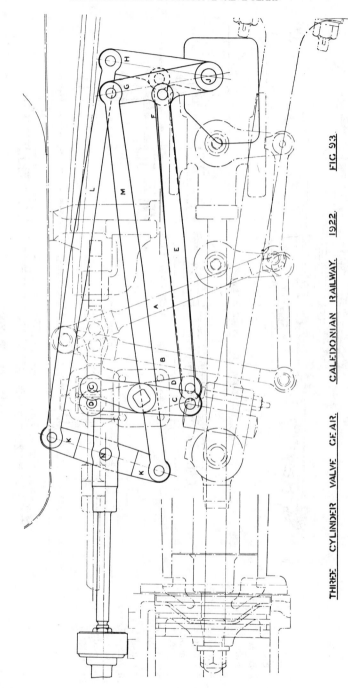

THREE CYLINDER VALVE GEAR.   CALEDONIAN RAILWAY.   1922.   FIG. 93.

levers respectively of Walschaerts valve gear for the outside cylinders. From the outside valve spindles two separate cross shafts operated the vertical rockers C and D, the lower arms of these rockers having each a driving rod E and F connected to the two levers G and H fulcrumed at J. The upper end of these levers were connected, one to the upper and one to the lower end of the equal armed lever K by the rods L and M; the centre of lever K was attached to the inside cylinder spindle at N. The centre valve spindle being therefore equally controlled by the motion received from the outside cylinder valve spindles, the main cranks being 120 degrees apart, as shown.*

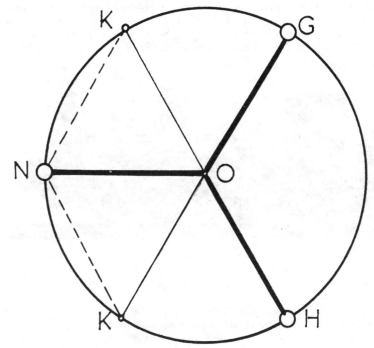

FIG. 93a.

This method of actuating the valves of the inside cylinder of a three cylinder locomotive having cranks 120 degrees apart may be described simply from Fig. 93a. The point G receives its motion from the left valve spindle and the point H from the right valve spindle, the point N being connected to the inside valve spindles. The movement of the inside valve spindle is that which would be produced by an imaginary eccentric ON obtained by the reversal of 180 degrees and composition of the movements of the

* For full description see Journal of Institution of Locomotive Engineers, No. 35, Paper No. 65.

supposed eccentrics OG and OH transferred to the positions K, K, as obtained by the double armed lever KK in Fig. 93.

Tappet Valve Gears.

Tappet gearing was first used on steam engines late in the 18th century. James Watt adopted round conical valves before the introduction of the slide valve. The first locomotive built in Scotland[51] by Messrs. Murdoch and Aitken, of Glasgow, in 1831, was fitted with tappet gearing operated by eccentrics. Also tappet gearing of the Edwards' type was applied to a locomotive on the St. Germains Railway, France, in 1840. In 1892 Durant and Lencauchez fitted valves of the Corliss type operated by Gooch link motion on engine No. 331 of the Paris Orleans Railway, this engine being built ten years previously by Sharp, Stewart and Co., of Manchester[52], twelve other engines of the same class being similarly

FIG. 94.
2-4-2 PASSENGER LOCOMOTIVE, No. 331, FITTED WITH CORLISS TYPE VALVE GEAR, PARIS-ORLEANS RLY.

fitted between 1892-95, two admission and two exhaust valves being used for each cylinder, this being illustrated by Fig. 94.

In the U.S.A. Young used an amended form of Stephenson's link motion to operate valves of the Corliss type. In this case only two valves were used for each cylinder, these being arranged at the top of each end of the cylinder, one valve serving for admission and exhaustion of steam from the cylinder.

Poppet Valve Gear.

Professor J. Stumpf, of Berlin, introduced a valve gear in 1909 for his uniflow cylinder engines in which the exhaust steam passed through slots in the middle of the cylinder, which were

---

51—Illustrated in "Elemental Locomotion." A. Gordon. 1832.
52—" Engineering," Vol. 58, page 291; " Locomotive Magazine," December 15th, 1914.

covered by a long trunk piston and uncovered by it when piston approached the dead points. Admission of steam to the cylinders was provided by two valves placed in separate chambers cast on the back and front cylinder covers. The inlet valves were operated by the valve rod of the usual type of Walschaerts' gear. The long trunk piston employed required a cylinder bore of about 508 longer than usual. The admission valves for this arrangement were mainly of the Lentz type. The N.E. Rly. 4.6.0 express goods locomotive No. 825 was fitted with uniflow cylinders in 1913, but about ten years later this engine was rebuilt with ordinary cylinders.

Fig. 95.
Beardmore-Caprotti Valve Gear for Two-Cylinder Locomotive.

In 1905 the Lentz poppet valve gear was adopted on a locomotive in Hanover. The Lentz gear consists of four similar valves for each cylinder. At the centre of the valve box a transverse tunnel envelopes the cam shaft to which an oscillating motion is imparted, usually by a modification of the Walschaerts gear. Later the Lentz system adopted a rotary drive owing to the reciprocating motion received from the Stephenson or Walschaerts gear giving practically the same steam distribution as engines fitted with piston valves.

Caprotti Gear.

The Caprotti poppet valve gear was first used on the Italian

State Railway in 1920, this gear also having four valves per cylinder, positioned vertically, one steam inlet and one exhaust valve at each end of the cylinder, as shown in Fig. 95. The longitudinal driving shaft (31) is between the frames and takes its motion from bevel wheels placed on a coupled axle (35). The front end of the shaft operates a bevel group rotating two cross shafts next to the cylinder casing and through them rotating the cam shaft (12) of the gear box.

Other forms of valve gear using poppet valves are the French Renaud gear, first used in 1928, and the Cossart rotary gear of 1933.

A.L.E. Gear.

In Fig. 96 is shown an arrangement of the rotary cam type poppet valve gear by the Associated Locomotive Equipment, Ltd. It will be seen that one cam box is mounted on each of the outside cylinders, each cam box being fitted with a cam shaft complete with cams for operating the admission and exhaust valves. The cams operate the valves through intermediate levers (63), as illustrated in Fig. 98. The intermediate levers are fitted with rollers (78) running on the cams and making contact at their upper ends with tappets bearing on the valve spindles (87). The main drive is by means of the standard " R.C." return crank gear box, Fig. 97, and a universal shafting. On this outside cylinder type of locomotive there is a complete main drive on each side of the engine, one set for each cam box.

Reversal of the locomotive is obtained by moving the cam shafts in a transverse direction to the valve spindles, this movement permits of any cam being brought into contact with the rollers in the intermediate levers as may be desired. A movement giving full travel to the cam shaft completely reverses the engine.

The pinions in each of the two cam boxes are coupled together by means of shafts and gears to the driver's reversing handle in the cab. The driver's control is very similar to that usually employed, and the index plate is marked, showing the actual position of the cam shafts. Each division mark indicates a definite cam and one turn of the driver's handle moves the cam shaft just sufficient to change from one cam to another. The catch of the reversing handle holds the gear in position for correct roller and cam contact, the cam in use being shown by the pointer and index plate.

Fig. 99 illustrates the cam shaft of an inside cylinder locomotive, also by the A.L.E., Ltd., the drive in this case being taken from one side of the engine and the reversing control being positioned on the other side.

Fluid Pressure Valve Gear.

Patents for hydraulic operated valves for steam engines have been taken out during the last 40 years, and 50 years ago David Joy experimented with a fluid pressure gear with a single eccentric. It was tried on a single wheel 2.2.2 type locomotive, No. 203,

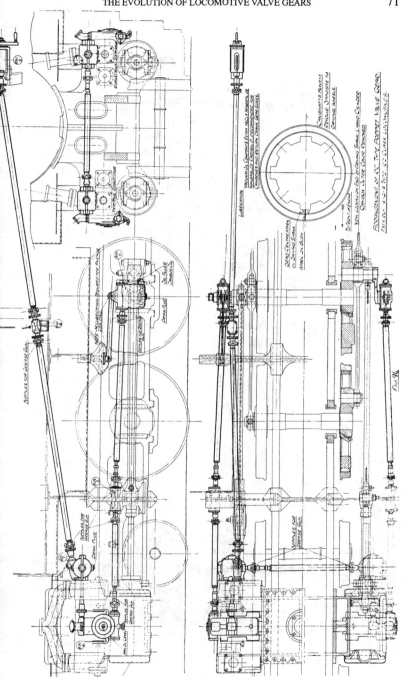

Fig. 96. Arrangement of R.C. Type Poppet Valve Gear. F.M.S. Railway, 4-6-4 Type, C2 Class Locomotives.

72 THE EVOLUTION OF LOCOMOTIVE VALVE GEARS

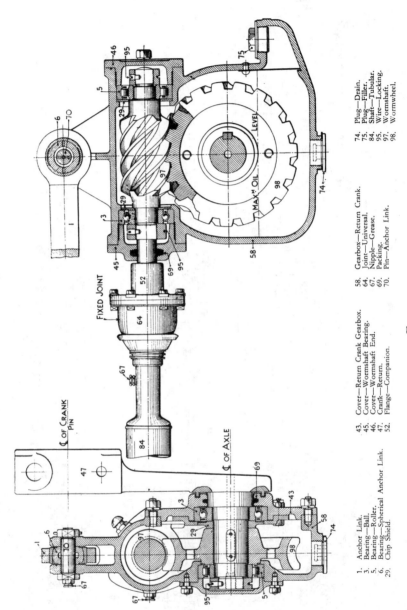

Fig. 97.—Deprit Valve Gear Return Crank Gear.

1. Anchor Link.
3. Bearing—Ball.
5. Bearing—Roller.
6. Bearing—Spherical Anchor Link.
29. Chip Shield.
43. Cover—Return Crank Gearbox.
45. Cover—Wormshaft Bearing.
46. Cover—Wormshaft End.
47. Crank—Return.
52. Flange—Companion.
58. Gearbox—Return Crank.
64. Joint—Universal.
67. Nipple—Grease.
69. Packing.
70. Pin—Anchor Link.
74. Plug—Drain.
75. Plug—Filler.
84. Shaft—Tubular.
95. Wire—Locking.
97. Wormshaft.
98. Wormwheel.

THE EVOLUTION OF LOCOMOTIVE VALVE GEARS 73

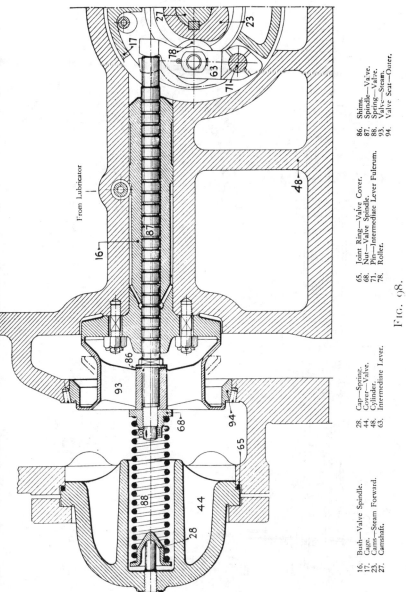

16. Bush—Valve Spindle.
17. Cage.
23. Cams—Steam Forward.
27. Camshaft.
28. Cap—Spring.
44. Cover—Valve.
48. Cylinder.
63. Intermediate Lever.
65. Joint Ring—Valve Cover.
68. Nut—Valve Spindle.
71. Pin—Intermediate Lever Fulcrum.
78. Roller.
86. Shims.
87. Spindle—Valve.
88. Spring—Valve.
93. Valve—Steam.
94. Valve Seat—Outer.

FIG. 98. POPPET VALVE GEAR, INLET STEAM VALVE.

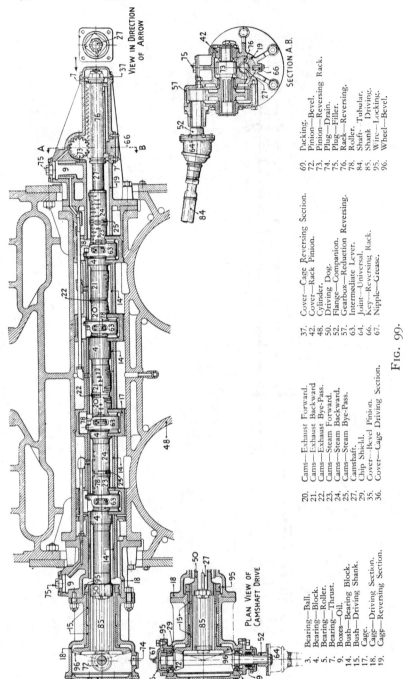

Fig. 99. "A.L.E.": "R.C." Poppet Valve Gear. Cam Shaft Inside Cylinder Engine.

3. Bearing—Ball.
4. Bearing—Block.
5. Bearing—Roller.
7. Bearing—Thrust.
9. Boxes, Oil.
14. Bush—Bearing Block.
15. Bush—Driving Shank.
17. Cage.
18. Cage—Driving Section.
19. Cage—Reversing Section.
20. Cams—Exhaust Forward.
21. Cams—Exhaust Backward.
22. Cams—Exhaust Bye-Pass.
23. Cams—Steam Forward.
24. Cams—Steam Backward.
25. Cams—Steam Bye-Pass.
27. Camshaft.
29. Chip Shield.
35. Cover—Bevel Pinion.
36. Cover—Cage Driving Section.
37. Cover—Cage Reversing Section.
42. Cover—Rack Pinion.
48. Cylinder.
50. Driving Dog.
52. Flange—Companion.
57. Gearbox—Reduction Reversing.
63. Intermediate Lever.
64. Joint—Universal.
66. Key—Reversing Rack.
67. Nipple—Grease.
69. Packing.
72. Pinion—Bevel.
73. Pinion—Reversing Rack.
74. Plug—Drain.
75. Plug—Filler.
76. Rack—Reversing.
78. Roller.
84. Shaft—Tubular.
85. Shank Driving.
95. Wire—Locking.
96. Wheel—Bevel.

THE EVOLUTION OF LOCOMOTIVE VALVE GEARS 75

"Sussex," of the L.B.S.C. Rly., in 1893.[53] There was an oblong slot in the eccentric, closed on either side, with a square shaft passing through it. The shaft moved up and down in the slot, for fore or back gear, according to the admission of oil above or below, and acting as a piston for the oil.

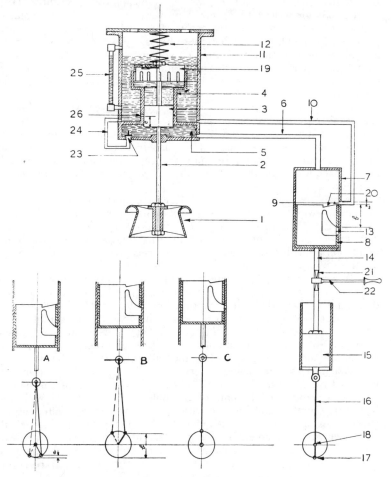

FIG. 100.   MEIER MATTERN OIL PRESSURE VALVE GEAR.

A recent oil pressure valve gear operating poppet valves is the Meier Mattern System and first applied on the Netherland Railways in 1929, the following being a short description of this novel form of valve gear.[54]

53—See "The Engineer," May 26th, 1893.
54—"Locomotive Magazine," Vol. XXXVIII, page 248.

In Fig. 100 the poppet valve (1) is required to be moved in sequence for admission of the steam to the cylinder and is connected by a rod (2) to the valve piston (3), known as the passive plunger, moving in the cylinder (4). The chamber (5) underneath is connected by a pipe line (6) to the cylinder (7) in which the pump piston (8), the active plunger, moves by means of an eccentric (17) on the driving axle. In order to prevent variations in the volume of the liquid by change of temperature or leakage the holes (9) are provided in the pump cylinder, these holes, by means of a pipe (10), are in communication with the equalising reservoir (11). This reservoir is partly filled with liquid and constitutes the highest part of the system, the holes in the cover giving free access to the atmosphere.

In the lowest position the piston (8), with each revolution of the crank axle, uncovers the ports (9) so that the equalising reservoir may supply any oil which might have leaked away or take up any surplus of oil when the volume has increased with an increase of temperature. Therefore, movement of the passive piston (3) and consequently of the valve will always commence at the same moment.

When the active piston (8) closes the port (9) the liquid displacement will begin and the passive piston will follow this displacement with a stroke inversely in proportion to the piston areas. The spring (12) keeps the piston in constant touch with the liquid and communicates the necessary acceleration to the moving parts on the downward stroke.

In order to obtain different admissions necessary in steam distribution the active piston (8) has been made hollow and the wall of the piston provided with one or more holes (13) of special type. The active piston can be turned about its axis by the handle (22) fastened on the square (21) of the rod (14). As soon as the holes (13) and (9) come in communication with each other by the upward stroke of the active piston (8) the oil pressure will be suddenly released into the equalising reservoir (11). The active piston will by this time have made a stroke " o." The turning of the handle (22) will bring a wider part of the hole (13) into touch with the hole (9), thereby reducing the stroke " b " and thereby effecting an earlier closing of the poppet valve (1) by the spring (12). With the downward movement of the passive piston (3) the quantity of oil used for the lifting movement will be forced through the pipe (6), cylinder (7) and pipe (10), back to the equalising reservoir (11).

The whole gear may be compared to some extent with the trip gear used with valve motions of stationary engines. The valve in this case being released hydraulically. An oil dash pot (19) assures a soft closing of the poppet valve.

In the same figure A, B and C shows three other positions of the active piston (8) during one revolution of the crankshaft. At A the port (9) is just closed by the upper side of the piston, the further upward movement bringing the oil under pressure, thereby starting the motion of the passive plunger (3). This is the point of admission which can be varied by the inclined recess (20) of the piston (9). At B the piston has traversed its stroke " b " showing the moment of release of pressure through the port (9) and the closing of the poppet valve by the acting of the spring (12). At C port (9) is fully open. During the return stroke of the piston (8) after having broken the connection between the ports (9) and (13), its further downward movement would create a vacuum in the space between the active and passive pistons, with a chance of air being drawn into the system. To prevent this a non-return valve (23) has been fitted, which, by means of the pipe (24) makes a connection with the equalising reservoir (11). During the return stroke of piston (8) the non-return valve will open, whereby a certain quantity of oil will be drawn in from the equalising reservoir, exactly equal to the displacement of the passive plunger (3) during its lift. In order to control any leakage of oil the equalising reservoir carried a gauge glass, as shown, and the equalising reservoir can be filled up as required.

The exhaust and compression are regulated in the same way as above, by the exhaust plungers and valves. Each cylinder having the usual two inlet and two exhaust valves per cylinder moved by their respective passive pistons.

Special provision has to be made for starting engines fitted with oil pressure valve gear, as all the valves will close as soon as the engine stops, owing to the drop of pressure in the control system. Starting is effected by means of an auxiliary steam cylinder, the piston of which is coupled to the plunger of the oil cylinder. As soon as the regulator is opened, steam enters the auxiliary cylinder, whereby the oil pressure will bring the liquid in the system under pressure, lifting the required valves, the active plungers being in the required position to give a free flow of oil to the respective passive plungers for lifting the valves.

The locomotive having two cranks at right angles to each other, the eccentrics for moving the oil pistons have been given an angle of advance of 45 degrees to their respective cranks. When running backwards one eccentric will have the same angle of advance with respect to the other crank. The steam distribution for the corresponding cylinder should therefore now be regulated by the same pump. This is accomplished by cross connecting the respective oil passages by means of a reversing cock actuated by a rod and handle from the footplate.

When the regulator handle which turns active plunger (8) about its axis is put in mid position, ports (13) and (9) are in constant connection with each other so that no pressure at all is attained and the poppet valves remain closed.

Bulleid Radial Valve Gear.

The most recent form of valve gear in this country is that used on the " Merchant Navy " class locomotives of the Southern Railway and patented by O. V. S. Bulleid.[55] This gear comes under the category of " Separate eccentric shaft gear " previously described and first used by Hackworth in 1827. The following is an abridged description taken from the original specification.

The object of the Bulleid gear is to provide an improved form of valve gear, which besides improving the steam distribution will overcome drawbacks associated with the usual form of valve gear.

Referring to Fig. 101 (a) is the locomotive frame and (b) a three-crank driving axle, and (c) the crank pin of the middle crank engaged by the big end (d) of the connecting rod (e). In the figure (h) is the centre of the three cylinders, the piston (g) of the outside cylinders being indicated by dotted lines. The valve motion, of the radial type, is not driven by return cranks or eccentrics on the driving axle, but by means of cranks or eccentrics on a secondary shaft (j) mounted in bearings (a) depending from the frame. The shaft (j) is driven by a sprocket and chain (i) from a countershaft K mounted on the frame at such an elevation as to be substantially level with the driving axle (b).

The countershaft K is driven by a sprocket and chain gearing (m) from the driving axle. Due to the level disposition of the countershaft (K) and its distance from the driving axle any disturbance of the valve timing, owing to the rise and fall of the driving axle in relation to the frame and under control of the spring suspension is negligible. The three cranks or eccentrics on the secondary shaft (j) are so disposed as to give the correct sequence of valve events in regard to the respective cylinders and the whole of the shaft (j) is within the distance between the engine frames.

The expansion or radius links (n) are mounted above the secondary shaft (j) and are rocked by their respective cranks or eccentric rods (o) connected with the respective cranks or eccentrics on the shaft (j). The upper end of each eccentric rod (o) is gudgeoned to an extension or bracket (n^1) extending from the respective link (n) so that the rods (o) occupy a substantially vertical position.

Combination levers (p) vertically arranged in the usual position are pivoted at (p^1) on the rear ends of respective short rods or plungers (q), the latter being connected with the valve rods Z. The position of the pivots (p^1) is governed by the valve steam admission according to whether outside or inside admission is adopted. The pivots (p^1) would be at the top of the levers (p) for inside admission, but with outside admission, as in figure, the pivots (p^1) are in an intermediate position.

Radius rods (r) are pivoted to the upper ends of respective combination levers (p). In the case of inside admission the radius rods (r) would be connected with levers (p) at an intermediate point. Each radius rod (r) extends reward beyond its respective expansion link (n) to which it is pivotally connected. The extensions (r^1) are connected at (S) with the reversing mechanism as shown.

55—Pat. No. 547156 of 17/8/42.

THE EVOLUTION OF LOCOMOTIVE VALVE GEARS 79

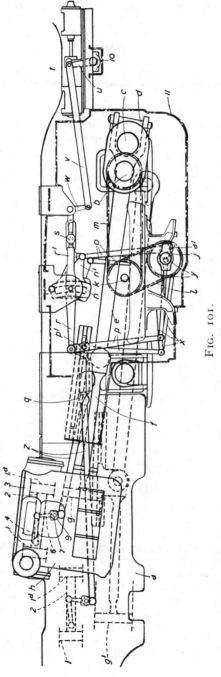

Fig. 101.
Bulleid Radial Valve Gear, 1942.

The lower end of each combination lever (p) is connected by a horizontal union link (x) with a pivot (y) of the big end of the respective vertically arranged eccentric rods (o). Alternately the links (x) may have direct connection with the crankpins of the respective eccentric rods (o).

The piston valve comprises two pistons (1) and (1^d) connected by two spaced webs (2) and operated within the cylinder (3). A horizontal link (4) may connect at one end with a pivot fixedly mounted between the webs (2) near to the piston head (1) and at the other end with the upper extremity of the arm (6) secured to the transverse rock shaft (7). A depending arm (9) is secured to the rock shaft (7) and is connected with the valve rod (z). The depending arms (9) of the rock shaft (7) belonging to the outside cylinders, as well as the arms (9) of the middle or inside cylinders are within the engine frame.

The location of the whole of the gearing parts within the frame permits of their total enclosure so that they may be protected and lubricated. The enclosed casing (11) is indicated by chain dotted lines, the covering being attached to the main frames. It will be observed that the casing (11) enclosed the big end of the inside connecting rod. Oil may be supplied to the various bearings by means of a pump driven from the secondary shaft (j) by chain or other gearing, the pump taking oil from the sump of casing (11).

Conclusion.

It will be generally accepted that for the period 1843-1910 the Howe-Stephenson link motion was the most popular form of valve gear, although the Gooch, Allan and Joy gears were, each in their turn, keen rivals. From 1910 the Walschaerts gear gradually gained popularity. Perhaps in this country this was due to the slow but sure decline of the inside cylinder locomotive, in addition to larger types introduced. R.C. poppet valve gears may be the future means of steam distribution in locomotive cylinders. However, from the maintenance aspect poppet valve gears will have difficulty in proving superior to the present narrow ring, long travel piston valve actuated by Walschaerts valve gear.

DISCUSSION.

Mr. H. Holcroft (M.) said that a feature of all radial valve gears, and also of valve gears with a fixed quadrant link, was that they gave constant lead; i.e. the amount of opening at the beginning of the stroke remained constant, but the angle at which the valve opened, of course, advanced with the notching up. In the case of a shifting link eccentric gear such as the Stephenson or Allan straight link motion, however, not only did the angle of advance increase with the notching up but so too did the amount by which the valve was opened, i.e. they gave variable lead. That was a most valuable asset of the Stephenson gear, in that the valve was well open for early cut-offs at the beginning of the stroke; and an engine with a Stephenson gear, if it was well designed, would give a larger mean effective pressure and horse-power for the same size of cylinder than other forms of gear operating reciprocating valves. By reduction of lead in full gear it gave greater accelerating power from rest.

All valve gears which derived their motion from cranks or eccentrics or from crossheads connected to cranks produced an approximation to harmonic motion, but it differed from the simple harmonic in that there were some minor harmonics compounded with it.

A simple harmonic motion could only be produced from cranks by gears having infinitely long rods, and in practical application the restricted length of rods and their angularities gave rise to minor harmonics of which those of any consequence were small in amplitude but of double the frequency of the main harmonic and at a phase of 90° to it, and these octaves had the effect of introducing variations in the movement of the valve which might be beneficial or otherwise. The length of connecting rod was another factor entering into the case. The merit of a gear of any type depended on its mechanical layout and proportions. There had always been a great deal of controversy about which gear was the best from all points of view, but apart from mechanical considerations there was little to choose between the numerous designs described by the Author, because they all generated more or less the same type of harmonic motion, notwithstanding wide differences in the mechanism employed.

It was largely a question of " the survival of the fittest " from the mechanical aspect, simplicity, cheapness of construction, adaptability and freedom from undue wear deciding the issue.

Mr. E. W. Marten (M.) described the Paper as a valuable record, in the form of a connected story, of the development of and research on valve gears over many years. He knew of no other publication, he said, in which so much information on the subject had been so well presented. The Paper indicated that a great deal of study had been given to the subject of steam distribution in the locomotive cylinder, but it was of interest to observe that the inherent disadvantage of the great majority of gears described was that they had fixed valve events, and it was not until towards the end of the Paper and therefore in comparatively recent years that one found efforts directed towards separating the events so as to make admission independent and unrelated to exhaust. No doubt the absence of competition from other forms of motive power partly relieved the drive towards improving cylinder efficiency, and, after all, the locomotive with a conventional type of valve gear did pull the train, and reasonably well at that; but what of the future?

It was in the future that he himself was chiefly interested, and he would like to ask whether the development of the reciprocating steam locomotive was to be prosecuted as vigorously as had that of, for example, the Diesel engine, in which case the improvement of cylinder performance offered the greatest scope, being that part of the locomotive where there was the greatest thermal loss. It was still a long way from the thermodynamic ideal obtainable of a maximum of some 20 per cent. With the probable further increase in steam pressures and superheat temperatures, if full advantage was to be taken of greater expansive working—or, in other words, if maximum use was to be made of the available heat drop in the steam—it would seem that valve gears of conventional type must eventually be displaced, particularly for high-speed operation or where efficiency and economy in working were of importance.

As an example of the difficulties encountered with the higher superheats, he knew of one series of locomotives where it had been found that the temperature of the superheat was such that the piston valve liners were badly scored with the result that it had been necessary to blank off some of the superheater elements. On the other hand, with those locomotives of the same class equipped with poppet valves, the absence of this trouble made it possible to take full advantage of the superior steam efficiency of high superheat. In discussing the present Paper, he did not want to press too much the subject of the poppet valve; but he was dealing with the future, and it was in that direction that development would probably be found to lie.

There was a brief reference in the Paper to the Caprotti gear. It was certainly the case that that gear was introduced in Italy about 1920, but the present British-Caprotti gear was in many detail respects very different from the original, and much had been and was being done in developing it to suit modern conditions of working and to avoid difficulties of maintenance. He alluded to that because of the Author's conclusion but he felt that provided it was well designed and of simple construction, a poppet valve gear should stand up to its work just as well as a conventional gear—in fact a cam-operated gear had much to its advantage on the score of maintenance because of its precision nature.

According to the Paper it took 35 years for the Walschaerts gear to be applied to a hundred or so locomotives—in other words, to become established—so at that rate there appeared to be a reasonable prospect of the poppet valve gear becoming universal during the lifetime of many present—assuming that the steam locomotive itself survived, which he thought was questionable unless its efficiency was increased by more progressive adoption of modern improvements.

Mr. W. Cyril Williams (M.) said the Paper was really an historical review and although of indisputable and lasting value to the *Journal* and must have entailed laborious research, no discussion was really invited. The Paper might have encouraged discussion if the Author had given his views on present or future valve gears. For instance, there was the Baker valve gear; about a week ago he had met an eminent American engineer who had since returned to the United States and asked him his opinion of this gear. His reply was: " You can cut it out; it is slowly disappearing." Is that so? If so, why?

Reference was made in the Paper to the maintenance aspect of the use of poppet valve gears, but the whole question of poppet valve gears would surely have to be very thoroughly studied as the speed of locomotives increased. In this connection he would like to mention that he had been most impressed, on a visit to Algiers just before war broke out, at the performance of the express Beyer-Garratt locomotives which were fitted with Cossart cam-operated vertical piston valves of the Nord type. They had been at work for several years, and ran with under 7 per cent. cut-off for more than one-third of the journey between Algiers and Oran, which is a distance of 262 miles. When one shut the regulator off one seemed to feel the whole engine leap forward, and high speeds were maintained for several miles with a remarkably slow drop. The performance, whether at speed or on 1 in 40 grade was excellent and efficient. He would admit that it was a " bag of tricks," but the interesting fact was that at the time of the landing in North Africa 28 out of 30 engines of this class with that gear were found working, although no spares had been received for two years.

Mr. E. C. Poultney, O.B.E, (M.) said the preparation of such a Paper was a monumental effort which would be most valuable as a work of reference when it appeared in the *Journal* of the Institution. Its compilation, he said, must have entailed a great deal of very hard work.

Mr. J. D. Rogers (M.) said the Paper was an encyclopædia on valve gears of the past, but personally, like Mr. Williams, he was more interested in the valve gears of the future. He had been surprised to learn how many valve gears had been invented in America, but, strangely enough, the only two to " get anywhere " had been invented on this side of the Atlantic—the Stephenson and the Walschaerts. The Walschaerts gear was invented in 1844, it was introduced in America by the B. & O.R.R. in 1904, and applied to the first Mallet built there. In 1906, the railway on which he served his time bought experimentally two engines fitted with this " new " gear. They did not know a name for it, but called it the " monkey motion " because as the engine ran it had the appearance of climbing or jumping. The beauty of it to them was that it got away from the use of fixed eccentrics, and, in spite of the many virtues of the Stephenson gear, it is of no use in America to-day, because the speed of the very large eccentrics would be so great that it would be impossible to lubricate them. The Walschaerts gear therefore serves the purpose, and some of the largest and fastest engines in America to-day, hauling the crack trains, are fitted with well-designed Walschaerts gear and piston valves.

So far as his own recent investigations went, he did not find that with any of the high-speed engines doing the maximum of 70 to 80 m.p.h. there had been any great trouble with the lubrication of piston valves, and he was surprised to hear that there had been trouble on this side of the Atlantic. The valves used in America are sometimes 12 inches or 14 inches in diameter, of light design, but it took a good deal of power to drive them, steam pressures being 250-300 lbs. per sq. inch.

He had used the Baker gear, and many other types, including the Southern gear with the fixed horizontal link; but that was just one that passed as an experiment. In his time the principal objection to the Baker gear was the pin connections, with fixed bearings. These collected lost motion quicker than Walschaerts gear, but now that it was on needle bearings it is doing a better job. Some of his friends told him that it was good only for freight and moderately fast passenger service, and that it did not give the distribution which was desired at high speed.

As far as poppet valves are concerned, there are only a few engines in America, built experimentally, which are using them. The Pennsylvania Railway has had several engines fitted, and some of the other railways. The most intriguing gear which had come under his observation is that fitted to the " Merchant Navy " class locomotives of the Southern Railway. When he first looked at the box and was told what was inside he was rather fearful that it might not do the job, but since he has travelled on the engine and seen the gear in the shops, he has changed his mind. He wondered whether some gear of that type was not the answer to the problem of driving piston valves—transferring the motion from the axle to the three-throw crank, which was, he thought, novel in valve gear construction.

Lt.-Col. K. Cantlie, R.E. (M.) also characterised the Paper as a monument of endeavour.

He said there were only two points where he could add to the Paper in any way. One was that the outstanding locomotives to be fitted with the Allan straight link motion were the L.N.W.R. " Jumbos," which many people regarded as among the most astonishing locomotives for their size ever built. The other was with regard to the fitting of Dendy Marshall valves to the L.N.W.R. locomotive No. 1361 " Prospero." The valves did not follow the fitting of the engine with four cylinders; the experiment was carried out to test Dendy Marshall valves.

He felt that in some ways it was a pity that the Author would not go one step further and indicate more fully than he had done why it was that the hundred and one gears described in the Paper failed in their various ways, why they were not perpetuated. Some of them looked excellent, but there must have been some fault or flaw in them. At the present date it was probably impossible to find out just what the trouble was, and in some cases, of course, the cause of the trouble was obvious; but in certain cases it might be of great interest if that information could be discovered.

One curse of valve gears lay in the up-and-down motion of the driving axle in the horns, and that, if he might say so, had been found to be the trouble with the Southern gear. Perhaps Mr. Rogers would agree with that. That also affected the Joy gear—not seriously, perhaps, but it could not have been a good effect. A difficulty with a cross-connected gear such as the Deeley, which looked very fine on paper, was that it could not be worked on one side, and if one side went the arrangement failed completely. That was a very important point which militated against cross-connected gears.

He thought it rather odd that the various types of flat valve, fitted with an expansion valve above or below them, were not perpetuated. The introduction of the link motion almost killed them, but before the link motion was introduced there was a very good reason for their use, in order to obtain expansive working of the steam. After the introduction of the Stephenson motion only a few efforts were made to continue their use. For working at very short cut-offs an expansion valve might have been worth the fitting, though history appeared to indicate that that was not the case. It might appear that the real reason was that the small steam pipes and small valve ports prevented working at very short cut-offs, but if that was not the reason then there should have been an advantage in putting in an expansion valve, because it would allow of freer exhaust, which was the point behind the use of poppet valves, etc.

One had only to listen to locomotives passing to hear the very large proportion in which the beat was definitely out; and, when it was remembered that the beat had to be 10 per cent. out before one could hear it, it would be realised that there was room for improvement. With the " Prince of Wales " engines on the L.N.W.R., the driver used to be blamed for wire-drawing the steam and working at half regulator until the engines were indicated and it was found that with a boiler pressure of 175 lb. per sq. in. the pressure in the cylinders was 170 lb. and the back pressure 220 lb. That was one reason why the drivers did not like working with short cut-offs.

The invention of valve gears appeared to be rather like the indicated horse-power of a locomotive; it rose very steeply when it started and gradually levelled off. After about 1860 it very much levelled off. That was an argument for the Stephenson and the Walschaerts being good enough, but, as had already been said, he did not think they would be good enough for the future.

Mr. D. C. Brown (M.) said that he felt it would aid the assimilation of the large amount of information which the Paper contained if the " high lights " of the changes in design from the functional point of view could be emphasised, either in a supplement to the present Paper or, if the author could be persuaded to crown his already considerable labours by a further Paper on the same subject.

Mr. F. L. Howard (A.M.) said his shed experience on the Southern Railway was that there was not a large amount of trouble with valve gears. They had to deal with the Walschaerts gear in its normal form and in the modified form on the 3-cylinder engines, and with the Stephenson working with the straight shaft and also through the rocking shaft with piston valves on superheater engines, and they did not meet with much trouble; maintenance was very light.

He did not think it could be said that the problem of lubrication had yet been solved. The old-fashioned slide valve had been a very wonderful fitting, and it gave good performance. They had a steam crane which had been in service for forty years, and it still had the original cast iron slide valves, and he did not think that they had worn 1/16th inch.

Mr. B. W. Anwell said it seemed that a large number of people had worked in a large number of different ways to try to attain the same ends. It struck him that at the present time locomotive engineers were not being original enough, but were being content with small modifications of already well-established principles and not looking ahead as far as they should. The type of valve used on the internal combustion engine withstood terrific loads and temperatures continuously, and in view of that it seemed that for locomotive work the poppet valve principle had not been sufficiently developed. The poppet valve would appear to have a great future in front of it, and if it proved possible to devise some improved method of controlling the admission with the poppet valve they would be getting a little nearer to the results which they sought to obtain. A good deal more could obviously be done on those lines with the development of alloy steels and other alloys which the war had produced, and once locomotive engineers were able to make use of those materials they might be a little nearer the solution of their problems.

Meanwhile, the old, radial types of gear did seem to be doing their job, and developments of them, and in particular the type used on the " Merchant Navy " class on the Southern Railway, with the oil bath, meant a definite step forward which would fill the gap until the poppet valve had been properly developed. He hoped that the reasons for the failure of earlier types of gear would be the subject of a Paper to the Institution in the future.

Mr. A. J. L. Winchester (M.) said he had recently been reading a publication by C. G. Wolfe which came out in the early years of the present century and described the various forms of link motion. The three main forms were described in detail, and in the case of the Allan straight link motion it was stated that this gear, when properly designed, gave the best distribution of steam of any gear so far used; and that would include what was practically the modern form of the Walschaerts gear, but not the poppet gears. If the present Author could throw a little light on that statement, and explain just what was meant by " the best steam distribution," it would be of interest.

Mr. O. V. S. Bulleid (President) associated himself with previous speakers in thanking the Author for his Paper, and expressed his regret that it had not been available twenty years ago. Searching through valve gears had always been a tedious and difficult task, but there would now be available in the Proceedings of the Institution a very concise and convenient history of the development of valve gears.

The earlier valve gears were efforts, he thought, to find a practical form of gear, and from the late 'eighties it might be said that the forms of gear were more or less settled to meet the then requirements; and the developments since had been very largely governed by practical conditions. Mr. Rogers had pointed out that in the large American engines eccentric gears of the Stephenson type were a practical impossibility. In Great Britain, which was still a " small engine " country, the Walschaerts and the Stephenson gears continued to be used.

As a matter of interest, he had compared the valve events of the first " austerity " engine built on the Southern Railway with those of the second " austerity " locomotive, produced by the Ministry of Supply. The Southern Railway engine used the Stephenson gear; the Ministry of Supply used the Walschaerts. When one looked at the figures, one had to admit that it was quite immaterial whether one used the one or the other; the events were both good, both engines did the work for which they were designed, and both stood up to their job. The Stephenson gear did not cause any trouble with lubrication. It was piston ring trouble rather than gear trouble which was generally experienced.

Going through the Paper, he could not help thinking of the past. He could just remember the engine on the G.N.R. which was fitted with the Marshall gear. He had been a youngster then, and at Doncaster they were all very interested in the engine fitted with this new contraption. Why it was taken off he did not know; it was probably a case of " Here's a foreigner; kill it! "

The next gear which amused him was the Cossart gear on the Nord Railway of France, which was not mentioned in the Paper. On one occasion when he happened to be in the Nord railway offices, he asked why anyone ever invented that gear, and the answer was amusing. It appeared that the Nord never found it necessary to borrow ideas from other people, and so M. Cossart was ordered, he believed by Baron Rothschild, to invent a valve gear. To maintain the tradition of not copying anything used anywhere else, therefore, the Cossart valve gear was evolved, and it was a very good valve gear. On one occasion Sir Nigel Gresley, who at one time had an unfortunate reputation in France—they called him the " Jonah," because whenever he travelled by the boat train from Calais it was almost certain that something would happen between Calais and Paris, and they even went so far as to suggest that they would prefer him to travel via Le Havre—was on the train and the engine failed at Creil, and the train had to be taken on to Paris by an 8-coupled tank engine fitted with the Cossart gear. He was very surprised to run into Paris behind this suburban tank engine at about two minutes under the normal time with a Pacific, and after that he was very interested in the Cossart gear, its performance having been rather startling.

Everyone was interested in the poppet gears, but they did not seem to have got as far as they should. They were certainly attractive, but there seemed to be something about them which prevented them making progress; what it was he did not pretend to know. It was a little shattering, however, to look at Fig. 99, and then turn to Fig. 92. There was a good deal of difference!

There was a tendency nowadays to talk about thousandths of an inch. He did not like thousandths; he preferred to talk of a sixty-fourth, or half a sixty-fourth. He liked the Irishman who referred to " a very small piece, about the back o' my nail." He understood that kind of engineering. When he heard his young men talk about thousandths he felt that the thing would probably seize or do something it should not do, whereas when one talked about sixty-fourths it would work. He had an uneasy feeling that when dealing with steam engines, mechanisms which were very " Dieselesque " in their make-up were perhaps a little out of place, and that if one stuck to that hideous contraption, the piston valve with rings, it might not look very nice, but it would very often run 50,000 miles without attention, which was not a bad performance. He thought it was really a question of applying sheer commonsense to the working of a machine which was not a precision machine, which never worked very long under constant load, and which was usually grossly over-powered. He thought that the question was really one of mechanical advantage pure and simple.

He was very grateful to the Author, and approved the suggestion that the Author should be invited to contribute another Paper, giving the pros and cons of the different gears, and giving, perhaps, a few more valve ellipses of the type used in the Paper to advertise the Baker gear—though why the Baker gear had been selected for that purpose he did not know.

WRITTEN CONTRIBUTIONS.

From **Mr. E. S. Cox** (M.): One cannot but marvel at the ingenuity displayed over the past 100 years in the devising of nearly as many different forms of locomotive valve gears. Leaving on one side the older types no longer in use, even the more modern existing forms are subjected to numerous variations, most of which seem to have been prompted by a theoretical rather than a practical approach. Any of the basic modern valve gears, Stephenson, Walschaerts or cam-driven poppet valve gear can be made to produce effective and economical steam distribution, granted correctly designed ports and passages, steam tight valves, and selection of the right valve events. While on paper it is possible to split hairs on the relative advantages of these gears from the distribution point of view, when fitted to engines and in daily service, no one is distinguishable above the others. As an example, the G.W. have produced long lap valve events with Stephenson gear, and it would be a bold man who could claim that a " Hall " class engine so fitted was more or less effective by virtue of its valve gear than an L.M.S. Class 5, mixed traffic engine with Walschaerts gear. Then again, there are a number of examples of the same class of engine fitted with both Walschaerts and poppet valve gears. Where the old gear did not allow the steam to be used expansively and gave rise to undue compression, and where the poppet gear was associated with improved passages and valve events, the latter, of course, showed up well. But where conditions are comparable no such clear advantage is to be seen. Most careful dynamometer car tests were made with the L.M.S. Standard 2-6-0 engines so fitted, and no difference in coal or water consumption could be discerned. Although both versions of this engine continue in active service there is still absolutely nothing to distinguish between them in this respect.

This verdict applies, of course, only to our present normal methods of designing and working engines. Where future accelerations demand super-power outputs, in conjunction with higher steam temperatures and pressures, then the finer points of distribution come into their own and tests on stationary plant may indicate a clear advantage of one gear over the other. But that is of the future so far as this country is concerned and for the present the case is, to say the least of it, not proven.

The most important present factor is, therefore, that of maintenance, and on this basis the Walschaerts gear is outstanding. There are 11 points of wear on an outside gear, all but two, those between curved link and block, and between valve rod and lifting block, being simple pin and bush assemblies. Moreover the arrangement of the leverages on this gear is such that cumulative wear at all but one of the pins is actually reduced in magnitude by the time it reaches the valve itself. This gives the great practical advantage that a really run down gear only affects the distribution in a relatively small degree.

Few parts of the locomotive give so little trouble in service or are so cheap and easy to repair as the Walschaerts valve gear. Changing a few bushes and a pin or two at each 50,000-60,000 mile shop repair is all that is required, while re-grinding of the curved and valve rod links is rarely necessary under 250,000 miles. In an endeavour to reduce even this small amount of maintenance, needle roller bearings have been recently introduced. They have not, however, so far produced any improved results because of the very difficult problem of sealing them against water and dust, within the small dimensional clearances which are possible.

It is just because the straightforward Walschaerts gear is so good that one marvels at the continuous efforts to supplant it, and far more effort and ingenuity seems to be exerted in this direction than towards other items of locomotive design which give infinitely more trouble, such, for example, as cracked plate frames and hot coupled axle boxes.

Reference is made in the Paper to conjugated gears for operating the inside cylinder of a 3-cylinder engine. The best known of these adds 8 pin joints to the existing two outside gears as against 8 pin joints plus two sets of sliding surfaces plus one eccentric if a full third Walschaerts gear had been employed. But a few lines set out on paper will show that the leverage of the 2 to 1 arrangement is such that a wear of x inches at the pins attached to the outside valve spindles becomes $3x$ inches of lost motion by the time the inside valve is reached and in addition, in the more usual form with the 2 to 1 levers in front of the cylinders expansion of the valve spindle where it passes through the steam chests is also multiplied by 3 by the time the conjugated motion reaches the centre valve. From a practical point of view, therefore, there is little which can be said in favour of such an arrangement. Where it involves some 14 pin bearings, as in the Caledonian Railway arrangement shown in Fig. 93, it ceases to be a practical proposition at all, and the four engines so fitted were notably unsuccessful. Tendency all over the world is to give 3-cylinder engines three independent gears.

Mr. C. F. Dendy Marshall (M.) wrote to say that the story about Humphrey Potter on Page 371 was now disbelieved by competent historians. See Dickinson's " Short History of the Steam Engine " (1939), p. 41. Page 392. It is quite true that American writers claim the invention of the link motion for W. T. James in 1832. But it may be added that there is no satisfactory contemporary evidence to support the claim. Page 408. Hackworth valve gear.

For many years he had been of the opinion that a form of that gear, properly designed, was most suitable for locomotives. But very few people fully understood it. Except on some agricultural engines, the eccentric rod had always been guided either by a radius rod, the length of which would be limited by considerations of space, or by a block with a straight slot. By employing a curved slot (the radius of which was much too long for a radius rod), the serious irregularities of the gear could be almost completely eliminated. These irregularities in its ordinary form were so great that marine engines fitted with it had even been made with double ports at one end of the cylinder in an attempt to cope with the situation. They were due to the obliquity of the eccentric rod, which was at a maximum in full gear and practically vanished in central gear, and to that of the rod running to the valve, which was very nearly the same in all gears. The irregularities in the distribution of all the gears of this type which had so far appeared were mainly due to the fact that the eccentric rod lay at right angles to the central line of the engine, and therefore had to be very short. In *The Engineer* for June 20th, 1913, he published a description and full analysis of a gear with the eccentric rod lying along the engine, at an angle of about 10° below the horizontal, with the end (beyond the block) connected to a bell crank, the other arm of which drove the valve. The eccentric rod was 7 feet long, which made its obliquity negligible; only leaving that of the link to the bell crank to be dealt with, which was done by curving the slot in the block to a radius of 3 ft. 6 ins. The arrangement was not altogether unlike that used afterwards in the Southern valve gear (Fig. 78), but in that case a radius rod was used, which could not be made long enough to give the required correction. An attempt was made in that direction by altering the angle of the bell crank from its natural value, but that expedient was only a palliative.

A few months ago it was stated in an article in the *Railway Magazine* that Engine No. 1850 on the Southern Railway was fitted with his valve gear, which was unsuccessful and was removed. The truth of the matter was that the gear tried on that engine was the invention of another Mr. Marshall.

Page 431. He asked to be allowed to add a short supplement to Mr. Shield's reference to his 4-cylinder system. It was far from being the failure that it might have been inferred to be from the fact that it was only used once. The " Prospero," originally built in 1907, ran for many years after having been converted to his system in 1915, without alteration. A report from the company, dated August, 1920, gave the coal consumption as 62.7 lb. per mile, and the average mileage between repairs in shops as 58,365; whereas the average figures for 49 similar engines not converted was 68.1 and 46,666. (Incidentally, there was an argument in favour of 4-cylinder engines.) Nevertheless, after the death of Mr. Bowen Cooke, the authorities decided not to give the system another trial.

THE AUTHOR'S REPLY.

With reference to remarks of Mr. Marten, he agreed that there was room for greater research being made with valve gears. Unfortunately his own practical experience was confined to Stephenson, Allan and Walschaerts gears. He also agreed with the President that working to fine limits in locomotive motion construction was asking for trouble; clearances on moving parts required to be ample to avoid heating.

In actual operation there was little difference between the Stephenson and Allan gears; both did their work satisfactorily and gave little or no trouble. Lt.-Col. Cantlie remarked on the reasons that the various gears were not perpetuated. Each new gear in its turn claimed some improvement over preceding gears, a claim that was generally not upheld in actual practice. To keep the Paper within reasonable limits only the transition from one gear to another was followed. The step from the early tappet gear to the loose reversing eccentric was a big advance in its time. The chief rival of the loose eccentric being the double-ended lever using one fixed eccentric. The coming of the double eccentric gab motion followed by the introduction of valve lap proved another advance in steam distribution. Shortly before the advent of the link motion the separate cut-off valve still further economised the steam consumption, however, at least in this country, the link motion superseded separate expansion or cut-off valves. Mr. H. Holcroft mentioned the advantages of the Stephenson gear. It has been claimed that the Walschaerts gear with its constant lead was a benefit. However, he was inclined to favour the Stephenson gear with its increasing lead toward mid-gear. A locomotive with a large amount of lead in full-gear was tardy at starting, and with the Walschaerts gear a satisfactory lead must be fixed upon to suit running conditions. The impetus to the Walschaerts gear since the first decade of the present century had been chiefly due to structural considerations and the necessity of employing large eccentrics with Stephenson's gear, to operate modern long travel piston valves.

In reply to *Mr. A. J. L. Winchester,* the best steam distribution, in his opinion, would be that giving the nearest approximate to the theoretical indicator diagram. That would necessitate a valve motion giving quick admission and cut-off in addition to full port opening from point of release to a constant point of compression. No reciprocating valve motion had yet attained that end. The modern poppet valve gears with only the events of admission and expansion variable and independent exhaust valve events was a move towards that ideal.

The President asked why the valve ellipse of the Baker gear was given in the Paper to the exclusion of all others; that was due to the Baker gear, as illustrated, being the most recent gear built in this country to which the valve ellipse was available. As regards the modern piston valve, that gave little trouble, and with proper attention to lubrication, engines usually ran between shopping periods without attention, running from 50,000 to 70,000 miles.

In reply to *Mr. Dendy Marshall,* the legend of Humphrey Potter was as probably untrue as otherwise, nevertheless one would like to believe the story of the boy playing on the engine room floor, while his engine was faithfully performing its duty with its sticks and strings, the valves opening and closing by the automatic action of engine, giving the first valve motion.

Mr. E. S. Cox referred to the elasticity of the levers in the conjugate valve gear applied to the inside cylinder of 3-cylinder locomotives. Trouble was experienced with the valve motion as given in Fig. 93, and subsequently a dashpot, as shown in Fig. 102, was fitted to absorb the inertia forces of the levers.

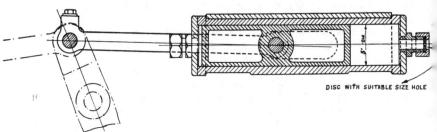

FIG. 102.
Dash pot fitted to motion in Fig. 93.